About IFPRI

The International Food Policy Research Institute (IFPRI), a CGIAR research center established in 1975, provides research-based policy solutions to sustainably reduce poverty and end hunger and malnutrition. IFPRI's strategic research aims to foster a climate-resilient and sustainable food supply; promote healthy diets and nutrition for all; build inclusive and efficient markets, trade systems, and food industries; transform agricultural and rural economies; and strengthen institutions and governance. Gender is integrated in all the Institute's work. Partnerships, communications, capacity strengthening, and data and knowledge management are essential components to translate IFPRI's research from action to impact. The Institute's regional and country programs play a critical role in responding to demand for food policy research and in delivering holistic support for country-led development. IFPRI collaborates with partners around the world.

www.ifpri.org

About CGIAR

CGIAR is a global research partnership for a food-secure future. CGIAR science is dedicated to reducing poverty, enhancing food and nutrition security, and improving natural resources and ecosystem services. Its research is carried out by 15 CGIAR Centers in close collaboration with hundreds of partners, including national and regional research institutes, civil society organizations, academia, development organizations, and the private sector.

www.cgiar.org

IFPRI

GLOBAL FOOD POLICY REPORT

2020

BUILDING INCLUSIVE FOOD SYSTEMS

A Peer-Reviewed Publication

RECOMMENDED CITATION: International Food Policy Research Institute. 2020. *2020 Global Food Policy Report: Building Inclusive Food Systems*. Washington, DC: International Food Policy Research Institute. https://doi.org/10.2499/9780896293670

International Food Policy Research Institute
1201 Eye Street, NW
Washington, DC 20005-3915 USA
www.ifpri.org

ISBN: 978-0-89629-367-0
DOI: https://doi.org/10.2499/9780896293670
Cataloging-in-Publication Data is available from the Library of Congress.

Photo credits
COVER: Chris de Bode/CGIAR.
CHAPTER IMAGES: p. 7 Sven Torfinn/Panos; p. 15 Joan Bardeletti/Panos; p. 29 Mikkel Ostergaard/Panos; p. 37 P. Lowe/CIMMYT; p. 47 Robin Hammond/Panos; p. 55 Giacomo Pirozzi/Panos; p. 66 Jim Holmes/Panos; p. 88 G.M.B. Akash/Panos.

BOOK LAYOUT: Jason Chow
EDITORIAL MANAGER: Pamela Stedman-Edwards
PROJECT COORDINATOR: Laura Zseleczky

Contents

Preface

The *2020 Global Food Policy Report* focuses on the need to build inclusive food systems, both to ensure that marginalized and vulnerable people enjoy the benefits and opportunities that food systems can bring and to support sustainable development. Inclusive food systems can help create better economic opportunities for poor people, mitigate climate change impacts for the most vulnerable, and spark innovation for the production and consumption of healthy foods. When this report went to print, the coronavirus outbreak posed a new threat. As the world battles this pandemic and as economies and livelihoods are disrupted, the poor and vulnerable are likely to suffer the most. Addressing the impacts of this shock–from supply chain and trade interruptions to severe unemployment to rising poverty levels–urgently requires effective, targeted social protection for the most vulnerable in the short term. For long term resilience, we must build inclusive food systems. This ninth annual report examines who is excluded in today's food systems and how to improve the terms on which individuals and groups participate so that everyone can reap their benefits.

In the report's opening chapter, former IFPRI director general Shenggen Fan and I review the importance of inclusion in food systems–what inclusion in food systems means; what benefits it could bring; the instruments, mechanisms, and policies needed for inclusion; and the next steps to achieve this vision. The chapters in the first section of the report provide more details and address specific aspects of inclusion in food systems. While smallholders cultivate the majority of farms in many countries, they often lack access to inputs, resources, and markets for profitable production. Young people face their own set of unique challenges to fully participate in food systems–particularly in Africa–including a lack of employment opportunities and limited access to resources such as land for agricultural production or financial capital for nonfarm enterprises. Women make significant contributions to food systems through food production and consumption, but heavy workloads and limited decision-making power and control over resources often leave them unable to make strategic life choices for themselves and their families. Refugees and people affected by conflict are especially dependent on agriculture, but pose significant challenges for integration into local food systems in their host communities. Finally, lessons from food system transformations at the national level provide policymakers and practitioners with recommendations to ensure that food systems transform in a healthy, sustainable, and equitable way.

The second section of the report focuses on specific regional aspects and efforts to make food systems more inclusive and provides a look ahead to 2020. A final section illustrates trends in key food policy indicators to provide a comprehensive overview of food policy and systems at country and regional levels.

Before concluding, it is important to express our great appreciation to Shenggen Fan, who launched this flagship report in 2011 and provided overall guidance for it over almost a decade. We hope this year's report encourages policymakers, business leaders, development practitioners, researchers, and the media to take action to build more inclusive food systems. All of us have a stake in food policies that include and benefit the world's poorest and most vulnerable people.

JOHAN SWINNEN
Director General

Acknowledgments

The *2020 Global Food Policy Report* was prepared under the overall leadership of a core team comprising Jamed Falik, Rajul Pandya-Lorch, Pamela Stedman-Edwards, Klaus von Grebmer, Sivan Yosef, and Laura Zseleczky. We would like to thank outgoing director general Shenggen Fan for guiding this flagship report over the last decade, as well as incoming director general Johan Swinnen for his support.

Text and data contributions were made by Fatma Abdelaziz, Akhter Ahmed, Kamiljon Akramov, Suresh Babu, Ousmane Badiane, Nienke Beintema, Clemens Breisinger, Inge Brouwer, Andrea Cattaneo, Nicola Cenacchi, Kevin Chen, Julia Collins, David Dawe, Alan de Brauw, Eugenio Díaz-Bonilla, Xinshen Diao, Paul Dorosh, Anjali Ganpule, Aulo Gelli, Kalle Hirvonen, John Hoddinott, Jarilkasin Ilyasov, Julius Jackson, Sally James, Bhushana Karandikar , Nadim Khouri, Avinash Kishore, Mengyao Li, Tsitsi Makombe, Hazel Malapit, Ian Masias, John McDermott, Ruth Meinzen-Dick, Purnima Menon, Valerie Mueller, Hung Nguyen, Alejandro Nin-Pratt, Allen Park, Valeria Piñeiro, Agnes Quisumbing, Abdul Wajid Rana, Delia Randolph, Shahidur Rashid, Roman Romashkin, Gracie Rosenbach, Devesh Roy, Marco V. Sánchez, Manika Sharma, Jenny Smart, Nandita Srivastava, James Thurlow, Peter Timmer, Rob Vos, Keith Wiebe, Sivan Yosef, and Laura Zseleczky.

Production of the report was led by Pamela Stedman-Edwards, with support from Jason Chow and Michael Go. Editorial assistance was provided by Tracy Brown and John Whitehead. Laura Zseleczky served as project coordinator.

We would like to thank IFPRI's donors, including all the donors who supported IFPRI's research through their contribution to the CGIAR Fund: https://cgiar.org/funders/.

The report underwent a peer review by IFPRI's Publications Review Committee, chaired by Gerald Shively.

CHAPTER 1

Reshaping Food Systems
The Imperative of Inclusion

SHENGGEN FAN AND JOHAN SWINNEN

Shenggen Fan is a senior chair professor, College of Economics and Management, China Agriculture University, Beijing, China, and former director general, International Food Policy Research Institute (IFPRI), Washington, DC, USA. **Johan Swinnen** is the director general, IFPRI.

KEY FINDINGS

- Inclusive food systems can help break the intergenerational cycle of poverty, hunger, and malnutrition.
- Including marginalized people in food systems can help them secure well-paying jobs and make gains in other areas that impact long-term livelihoods, such as education.
- A value chain framework is key to designing inclusive food systems—from improving farmers' access to resources and information to creating off-farm jobs and enterprises in the midstream of the chain.
- Recent innovations such as mobile phone technologies offer opportunities for marginalized and excluded populations to access information and services, and to participate all along the food value chain.
- Education is a major driver of inclusion, increasing lifelong income and improving nutrition, health, civic engagement, and gender equality.
- Marginalized people should be empowered to make strategic choices within food systems and have a voice in holding governments accountable for delivery of inclusive food systems.

RECOMMENDATIONS

- Address inclusion at the global policy level, using awareness of inequality to spur discussion of the need for large-scale investments in research and programming to build inclusive food systems.
- Take action at the national level so that the local context—including the status of specific populations, economic structure, and cultural norms—can be taken into account in shaping inclusive food systems and improving diets.
- Tailor food system policies so that they create opportunities for marginalized people while addressing key challenges such as unhealthy diets and climate change.
- Identify the needs of marginalized people early on, and give them a voice in research and policy- and program-design processes.
- Recognize the contributions that excluded people already make to food systems with their time and labor through policies that empower them to secure more equal benefits.

Our food systems are at a critical juncture. The challenges the world faces in feeding a growing population may seem familiar, but their scale and the pace of change taking place in global, regional, national, and local food systems are unprecedented. After making significant strides in reducing hunger during the past decades, our progress has slowed and, by some measures, has been reversed: in 2018, 820 million people were projected to be hungry—a figure that has climbed for three consecutive years—and a quarter of the global population faced moderate to extreme food insecurity.[1] Overweight and obesity are rising in almost every country, and progress on key nutrition indicators such as child stunting and exclusive breastfeeding has lagged, putting the Sustainable Development Goal (SDG) of zero hunger by 2030 seemingly out of reach. But just as critical, global inequality persists. Even as emerging economies and developing countries have continued to grow, albeit slowly, their citizens do not enjoy equal access to resources associated with economic development and a better quality of life. As a result, the world's poorest and most vulnerable are likely to bear the brunt of shocks, including the deadly global outbreak of the novel coronavirus in late 2019 and early 2020, that disrupt livelihoods and food systems. Hunger and malnutrition are likely to rise in 2020 as the pandemic impacts all aspects of our food systems. In the short term, targeted programs are needed to protect children, women, and other vulnerable population groups. To reduce the impact of such shocks in the long term, we must build more resilient and inclusive food systems. It is currently too difficult for the world's poorest and most vulnerable to enjoy these systems' outcomes, such as affordable, safe, and nutritious foods, or to share fairly in their economic benefits.

Food systems have also not yet addressed other looming challenges. Agriculture, for example, accounts for 24 percent of greenhouse gas emissions,[2] and while poor people are most vulnerable to the effects of climate change, they still have little power over the nature and speed of mitigation and adaptation actions. Urbanization is speeding up—most of the world already lives in urban areas, with many rural people, especially youth, migrating to small and midsized

The authors thank Sivan Yosef, senior program manager, Director General's Office, IFPRI, for her support in writing this chapter.

towns and larger cities. National and local food systems have done little to integrate these populations so that they can seize employment opportunities all along the food value chain. This leaves many of them trapped in low-productivity sectors.

But for all their flaws, food systems at all levels can also offer a much needed solution to these immediate challenges, especially when they integrate historically excluded people at all stages of the agrifood value chain and involve them in the decision-making processes that shape the programs, policies, and investments affecting their day-to-day lives. Inclusive food systems can help mitigate climate change impacts for the most vulnerable and also foster innovation to achieve climate-smart agriculture. They can create better economic opportunities for poor people, who are most often employed in agriculture and other food-related sectors, thereby reducing hunger and poverty. They can spark innovation, such as reorienting production and consumption toward healthy foods, thus improving nutrition for both producers and consumers and boosting the incomes of producers. Finally, and equally important, inclusive food systems can help build a sense of community and a personal stake in national success, possibly contributing to political stability.

Beyond the usefulness of inclusive food systems in addressing the world's most pressing challenges, inclusiveness is a moral imperative. Most people want to live in a world free of poverty, hunger, and malnutrition, and the world has committed to this ambition through the SDGs. Achieving this vision requires that particularly downtrodden groups in society reap greater benefits from the food systems with which they interact. These groups are diverse and also overlap in identity. Smallholders cultivate the majority of farm units in many countries but produce only a third of the total value of the agricultural food supply,[3] due to their lack of access to nonstaple seeds, land, and profitable markets (see Chapter 2). Similarly, despite their substantial contributions to agricultural production and household food and nutrition security, women face heavy workloads and have less decision-making power than men. They also control fewer resources within their households and communities (see Chapter 4).[4] Youth are also marginalized in many countries, lacking sufficient employment opportunities, land if they choose to stay in agriculture, and financial capital if they attempt to enter the rural nonfarm economy. These issues are particularly acute in Africa, which will see 30 million youth entering the workforce annually by 2050 (see Chapter 3). Conflict-affected people and refugees, who may have fled their homes due to political, ethnic, or religious strife or climate-induced weather shocks, are mostly rural and dependent on agriculture.[5] Refugees typically stay in their new

BOX 1 WHAT ARE FOOD SYSTEMS?

Food systems are the sum of actors and interactions along the food value chain–from input supply and production of crops, livestock, fish, and other agricultural commodities to transportation, processing, retailing, wholesaling, and preparation of foods to consumption and disposal. Food systems also include the enabling policy environments and cultural norms around food.

Food systems provide basic sustenance in terms of meeting populations' minimum caloric needs and affect nutrition, positively or negatively, through crop health, dietary diversity, and impacts on human health and the environment. Food systems also provide livelihoods for a sizable share of the global population, through agricultural labor and nonfarm jobs in other segments of the food value chain. The income garnered from these jobs can be used to purchase a wide array of healthy foods, send children to school, purchase health services and medications, and more. At the macro level, food systems power local and national economies, shaped in part by governance, trade, and investment at the global level.

Ideal food systems would be nutrition-, health-, and safety-driven, productive and efficient (and thus able to deliver affordable food), environmentally sustainable and climate-smart, and inclusive. But to realize this vision, continued investments must be made in agricultural research and development and technological innovations, paving the way for programs and policies that are based on sound evidence.

locations for long periods of time, posing serious challenges for the creation of livelihood opportunities and integration into local food systems (see Chapter 5). Other examples of excluded people in the world today include the elderly, lower castes, religious and ethnic minorities, and people with disabilities. Each of these groups faces a unique set of challenges.

The world is transforming at a breathtaking pace, and food systems must evolve quickly to meet growing and changing demand. Innovation is essential to transforming food systems so that they bring a wide range of benefits to all people. As we modernize food systems to make them climate smart, healthy, and sustainable, we must also strive to make them inclusive.

WHAT IS INCLUSION IN FOOD SYSTEMS?

Inclusive food systems reach, benefit, and empower all people, especially socially and economically disadvantaged individuals and groups in society.[6] Inclusive food systems reach vulnerable people by way of reducing barriers that currently prevent them from participating in food system activities, for example, by enabling them to gain the skills needed to work within evolving food value chains. The benefits of inclusive food systems, such as access to affordable, safe, and nutritious foods, extend to all people, including poor consumers. Inclusive food systems also allow everyone to share fairly in their economic benefits—young people and women can find remunerative jobs and participate in activities that add value to foods, and smallholders have access to food and agricultural markets. Ultimately, this means a more participatory way of shaping food systems.

Inclusive food systems empower people to make strategic life choices, such as when they increase women's decision-making power within their households. They give marginalized people a voice in local food policies that affect their daily lives and open leadership opportunities at the local, national, and global levels. Inclusive, participatory decision-making can contribute to improved governance, and can yield legislation that is more relevant to the issues facing poor and underrepresented people and has more local buy-in. Creating climate-smart policies, for example, may be best done by poor farmers who possess a deep understanding of the local context, from trade-offs between production and environmental health to assessment of risks.

Inclusion is an action-oriented concept that is closely tied to the social goals of equality (fair and equal treatment) and diversity (for example, an appreciation of different ethnicities, religions, genders, and disabilities). Promoting inclusion is a practical means for individuals, private firms, institutions, policymakers, and governments to ensure that vulnerable people have access to services and opportunities. The quality of these benefits is important too—for example, women provide much of the labor in food systems, yet often have limited land tenure rights. In this sense, inclusion builds upon the notion of equity—giving everyone what they need to live healthy and fulfilling lives.

WHAT ARE THE BENEFITS OF CREATING INCLUSIVE FOOD SYSTEMS?

Reshaping food systems to be inclusive of poor and vulnerable people is a moral imperative. But the policy world is complex, with policymakers constantly weighing short- and long-term costs, benefits, and risks, the interests of wide and disparate groups of people and institutions, and their own political survival. Against this backdrop, it is useful to identify the wide-ranging economic and human development benefits associated with inclusive food systems.

In low-income countries, the agrifood sector supports many people's livelihoods. In 2019, 63 percent of people in low-income countries were employed in agriculture.[7] Better integrating marginalized people into national food systems, by linking subsistence-level farmers to markets or incentivizing farming households to move out of agriculture and into other areas of the food value chain, is perhaps the most effective way to achieve inclusive economic growth. By increasing household income, inclusion can help reduce absolute poverty and help poor households access other services and benefits closely associated with poverty reduction, such as education, nutrition, water and sanitation, and healthcare. Inclusive food systems can also break the intergenerational cycle of poverty, hunger, and malnutrition. This potential impact can be seen most clearly in the latest literature on women's empowerment. When women have increased decision-making power over household income or more control over assets, agricultural productivity rises and household food security, diet quality, and maternal and child nutrition improve.[8]

Inclusion can also help reduce global and national-level inequalities. Global inequality has spiked since 1980. Its growth has not been steady, exhibiting a slight decline after 2000, but it remains at very high levels. National-level inequality has risen in nearly all global regions, albeit at different speeds, and has been notably high in the Middle East, Africa south of the Sahara, and Latin America.[9] Including marginalized people in food systems can help them to not only secure well-paying jobs but also make gains in other areas, such as education, equalizing human development as well.[10] A well-educated populace can better advocate for pro-poor policies, such as progressive taxation and open access to financial information, as well as reducing tax evasion and corruption, which are key drivers of inequality.

Inclusion also supports the proliferation of diverse ideas about how to improve processes and strategies not only within local food systems but also the global food system. Research on the relationship between inclusion and innovation in private sector entities has shown that diverse work teams develop more innovative ideas. Inclusion can also boost profits: a recent study of more than a thousand companies in 35 countries found that, when the national culture valued diversity, gender diversity was associated with more financially productive enterprises.[11]

Inclusion may contribute to political stability. The Political Instability Task Force, a research partnership that forecasts political instability, focuses on four key factors that put countries at risk of instability: high infant mortality; unstable neighboring countries; weakly institutionalized democracy; and the exclusion of minority, ethnic, or religious groups. Thus social and economic polarization are key drivers of citizen dissent.[12] But when all citizens feel that they can obtain a good livelihood, access high-quality services regardless of their identity or geographic location, and have a voice in the way decisions are made, they also feel invested in their countries' and communities' future.

WHAT ARE THE INSTRUMENTS, MECHANISMS, AND POLICIES FOR INCLUSION?

As food systems evolve, many different types of mechanisms can be put into place to ensure that they reach, benefit, and empower vulnerable people. Many of these actions should ideally be implemented at the national level, so that the local context, including the status of specific populations, economic structure, and cultural norms, can be taken into account (see Chapter 6).

VALUE CHAINS

Some of the most relevant actions that can be taken to redesign food systems are those that use a value-chain framework. Such an approach can focus on the beginning of the chain, improving excluded people's access to natural resources such as land (through land tenure security, an especially salient issue for women and the landless), water, or seeds. Integrated agriculture and nutrition interventions such as biofortification show particular promise for bringing smallholders into "healthy" value chains that promote a nutritious diet, from seeds to consumption. For example, a recent evaluation of a HarvestPlus project that distributed biofortified orange sweet potato vines to households in Mozambique to grow for both own-consumption and selling found that vitamin A intake remained higher among children in participating households than nonparticipating households three years after the project ended.[13]

A value chain approach can also zoom in on the "hidden middle" of the food value chain, including processing, distribution, and services (see Chapter 2), where the potential for creating enterprises and jobs is greatest. As food value chains become longer and more complex in response to urban demand, there will be a critical need for ensuring food safety and quality through regulation, certification, and inspection as well as innovations for cold storage and transportation. Investing in the institutions and infrastructure needed to serve urban markets represents a win-win for job creation, consumer health, and developing countries' exports.[14]

SOCIAL PROTECTION

Social protection can safeguard food and nutrition security for marginalized people. In desperate situations, food and cash transfers can fulfill basic caloric needs and prevent malnutrition. An example of the power of social protection amid crisis, cash transfers in war-torn Yemen prompted households to increase food purchases by 17 percent and spend the money on nutrient-rich vegetables, fruits, and animal-sourced foods such as milk and eggs.[15] Social protection can also free up resources to use for healthcare, education, and other services, or to allow poor people to take up more profitable, non-farm entrepreneurial ventures within the food system. Ethiopia's Productive Safety Net Program, for example, which provides cash or food transfers in conjunction with public works and livelihood support programs, has increased participants' probability of engaging in non-farm activities by 5 to 7 percentage points.[16]

Conditional transfer programs can also promote improvements in nutrition, school attendance, or rural employment and improvements in agricultural production, such as homestead gardening. Some school feeding programs turn to smallholder farmers to supply locally grown, nutritious foods, thus creating a local, healthy food system that includes poor farmers. Economic transfers also have a host of other potential benefits for excluded groups. For example, cash transfer programs have been shown to decrease intimate partner violence in low- and middle-income countries, presumably by increasing economic security and emotional well-being.[17]

EDUCATION AND INFORMATION

Education is perhaps the greatest driver of inclusion. Not only does education increase lifelong income, breaking the cycle of poverty, but it also improves nutrition, health, civic engagement, and gender equality. Education in the form of vocational training can also create a well-trained labor force that can seize opportunities in higher-productivity food-related sectors, a prospect that is especially beneficial for youth.

Facilitating the transfer of knowledge and reducing information asymmetries between the rich and poor, urban and rural people, men and women, and so on, is another key driver of inclusion. New technological innovations, including mobile technologies, are creating numerous opportunities for poor and vulnerable people, who now have at their fingertips information on agriculture, markets, and nutrition. Up-to-date information about prevailing market prices, for example, can help rural farmers get the best price for their crops, and information about the budget of a local government can help citizens press for accountability on spending. But for information to be useful, it must be easily understandable and relevant to citizens, and they must be able to act upon the information.[18] Inclusive governance processes must be in place so that citizens can translate information into improved services.

GOVERNANCE AND LEADERSHIP

It is not just the outcomes of food systems that need to be inclusive–marginalized people should also be included in the process of designing food-system-related policies and programs, and have a voice in monitoring, evaluating, and holding institutions and people in power accountable for the delivery of high-quality jobs and services. Scorecards, for example, can help boost accountability by tracking the inclusiveness of food systems according to indicators related to nutrition, employment, climate change, and more. IFPRI and partners are currently conceptualizing a Global Food Systems Index, which would monitor and track progress toward a desired food system; inclusiveness could be one measure of success.

Leadership also matters. Women, for example, are severely underrepresented at senior levels within international organizations, donor agencies, national-level political leadership, education management, and large businesses. When marginalized people are in positions of power, they can integrate the interests of excluded populations into policy and program design and implementation. Indeed, private sector research has shown that diverse employees are more likely to have common experiences with their company's end users.

WHAT ARE THE NEXT STEPS FOR INCLUSIVE FOOD SYSTEMS?

Policymakers, researchers, and program designers can take concrete steps today to create inclusive food systems, with numerous benefits for both marginalized groups and the larger society of which they are a part.

INTEGRATE INCLUSION INTO THE GLOBAL AGENDA. Inclusion needs to be addressed at the global policy level. Currently, the only international standards for inclusion can be found within the Charter of the United Nations. Global policy forums can seize upon the increased awareness of inequality to discuss the need for inclusive food systems, making way for large-scale investments in research and programming.

SEIZE OPPORTUNITIES. As challenges mount, food system policies can be tailored to both address the hurdles and seize new opportunities. The hurdles associated with urbanization, such as longer food chains, higher consumption of processed foods, and underemployment, can be tackled by ensuring that marginalized people have the skills to participate in forward-looking innovations such as healthy value chains and the information and communications technology revolution. Territorial approaches, such as agro-industrial parks and incubators, can cost-efficiently provide rural areas with services and support development of the middle segments of the agrifood value chain.[19] Addressing climate change in part through the development of renewable energy systems may also generate new employment opportunities. These are just a few examples of how inclusion can facilitate innovation.

IDENTIFY NEEDS EARLY ON. Researchers can identify marginalized people's needs and priorities in early stages of food systems research. Such an effort could yield important insights, such as whether some smallholders are well suited to move up in agriculture (from small to mid- or large-scale farming) or should move out of agriculture to other parts of the food value chain or other sectors. Inclusion in this process can be achieved by using participatory research methods and even working to create research career paths for historically underrepresented people. Policymakers and program designers can similarly ensure that excluded people are represented in all stages of policy and intervention design, implementation, and evaluation, as well as in decision-making institutions.

RECOGNIZE EXISTING CONTRIBUTIONS. In almost all cases, excluded people already make immense contributions to food systems in terms of their time, workload, and the health risks taken on. Their participation should be recognized in policies that secure them more equal benefits. For example, fair contracts can enhance the negotiating power of small and often informal actors along the value chain and improvements in land tenure security can help the poor build assets.

Inclusion is not a panacea. It is one of a number of innovative remedies to food systems that have in many ways failed poor and marginalized people. We must reinvent these systems, and do so now. Challenges such as climate change, the double burden of malnutrition, and the coronavirus pandemic are already exacting a heavy toll, especially among the most dis-advantaged populations. But if we build on innovations and continue to pioneer new ideas, we can design food systems that are inclusive, climate smart, and sustainable, and we can provide healthy diets for everyone. The future well-being of all the world's citizens depends on it.

"Reshaping food systems to be inclusive of poor and vulnerable people is a moral imperative."

CHAPTER 2

Smallholders and Rural People
Making Food System Value Chains Inclusive

ROB VOS AND ANDREA CATTANEO

Rob Vos is the director of the Markets, Trade, and Institutions Division, International Food Policy Research Institute, Washington, DC, USA. **Andrea Cattaneo** is a senior economist, Food and Agriculture Organization of the United Nations, Rome, Italy.

KEY FINDINGS

- Propelled by urbanization, rising incomes, and changing diets, food markets are expanding in Africa and South Asia, creating enormous potential for job and income opportunities along food supply chains.
- Small and medium-sized enterprises have proliferated in storage, logistics, transportation, and wholesale and retail distribution to meet growing rural and urban food demands. This so-called quiet revolution appears to be taking place out of sight of policymakers, leaving much of the potential for inclusive value-chain development untapped.
- Smallholders often struggle to connect with actors in the middle of the food supply chain as a result of limited access to land and inputs and lack of capacity to scale up or implement new practices to meet quality requirements.
- Lack of infrastructure and skills is holding back the development of food supply chains in low-income Africa and Asia, especially where the potential is greatest: in small towns and intermediate cities near rural farmlands.

RECOMMENDATIONS

- Promote inclusive food supply chain development by leveraging the transformations already taking place in downstream food supply chains, particularly the expansion of small and medium-sized enterprises and growth of off-farm employment.
- Catalyze investments that strengthen food supply links so that smallholders have greater market access and food transporters, distributors, processors, and retailers can thrive. Governments should create an enabling environment for agrifood businesses by providing basic infrastructure, creating the right market incentives, promoting inclusive agribusiness models, and supporting information and communications technology use that fosters inclusive value chains.
- Enable smallholder engagement in dynamic food supply chains by addressing issues that hinder participation. Policies and regulatory frameworks should ensure land tenure security, access to credit, training and technical assistance, and resilience-enhancing social protection.
- Make much greater investments in data collection and analysis across the entire food system, particularly for the "hidden middle," to underpin policies for inclusive value chains.

Reducing poverty and ending hunger depend on making progress in rural areas, where most of the world's poor and undernourished live. Since the 1990s, rural transformation in many of the poorest countries has helped more than 750 million people move out of extreme poverty.[1] Boosting smallholder productivity and incomes and creating off-farm employment by developing the downstream segments of food value chains could be keys to achieving the same for those who remain behind. Agrifood system transformation is therefore critical for greater inclusion of smallholder households and other rural people. This chapter outlines a range of policy options to leverage this enormous untapped potential.

Industrialization, the main driver of past structural transformations, is lagging in most countries of Africa south of the Sahara and South Asia. In these poorest regions, rapid urbanization is not being matched by commensurate growth in employment and income opportunities in manufacturing and modern service sectors. As a result, most workers exiting low-productivity agriculture are moving into low-productivity informal services, usually in urban or peri-urban areas.

The benefits of this type of transformation are modest. Since the 1990s, poverty rates in Africa have declined little, while the absolute number of poor has risen.[2] Poor rural Africans migrating to cities are more likely to join the masses of urban poor than to find a pathway out of poverty. A similar dynamic is occurring in South Asia, where the rural poor are more likely to escape poverty by staying in rural areas than by moving to cities.[3] Growing demographic pressures will exacerbate these challenges: by 2030, the combined population of Africa and Asia is projected to increase from 5.6 billion to more than 6.6 billion. In this context, the world's 510 million smallholder farmers (those farming under 2 hectares), whose prospects for finding better jobs are already bleak, risk falling even farther behind.

Despite their precarious position, smallholders play a large role in the food system. According to a recent FAO study, they produce roughly 36 percent of the value of the world's agricultural food supply.[4] In China and India, the shares are significantly higher, at 80 and 50 percent, respectively. In Africa south of the Sahara and South Asia (excluding India), smallholdings comprise 70 to 75 percent of farm units, but they generate just 35 to 40 percent of the primary production value of the domestic food sector—substantial, but far less than often claimed (Box 1).[5] Limited access to land and inputs and concentration on production of inexpensive staple crops explain the disproportionately low share of agrifood value added earned by small-scale farmers. Their more inclusive participation in food-sector growth therefore has significant potential to reduce poverty and improve livelihoods.

Growing urban markets will continue to be the main drivers of agrifood sector expansion, including in Africa and Asia. Urban populations already consume up to 70 percent of the world's food supply, even in countries with large rural populations.[6] Income growth is driving a dietary transition, as urban consumers shift consumption from staple cereals toward high-value fish, meat, eggs, dairy products, fruits, vegetables,

BOX 1 SMALLHOLDERS IN THE FOOD SYSTEM

Agriculture is the predominant economic activity in rural areas of developing countries, and smallholders make up the largest share of farmers. About 1.5 billion people, often poor, live in smallholder households. What constitutes a small farm varies within and across countries, depending on socioeconomic and agro-ecological conditions, but a threshold of 2 hectares is often used to define "small." Worldwide, 510 million farms (84 percent of an estimated total of 608 million farms) are less than 2 hectares, while 70 percent of farms cultivate less than 1 hectare. Small farms account for only 11 percent of the world's farmland, but in poorer countries, small farms occupy a much larger share of the land—almost 40 percent of farmland in Africa and South Asia. Smallholders there generate 35 to 40 percent of the primary production value of domestic food production—a significant share, but far less than often claimed (Figure B1). This should not be surprising, given smallholders' limited access to land and inputs and dedication to the production of generally low-priced staple crops.

FIGURE B1 Smallholder share in value of primary food production

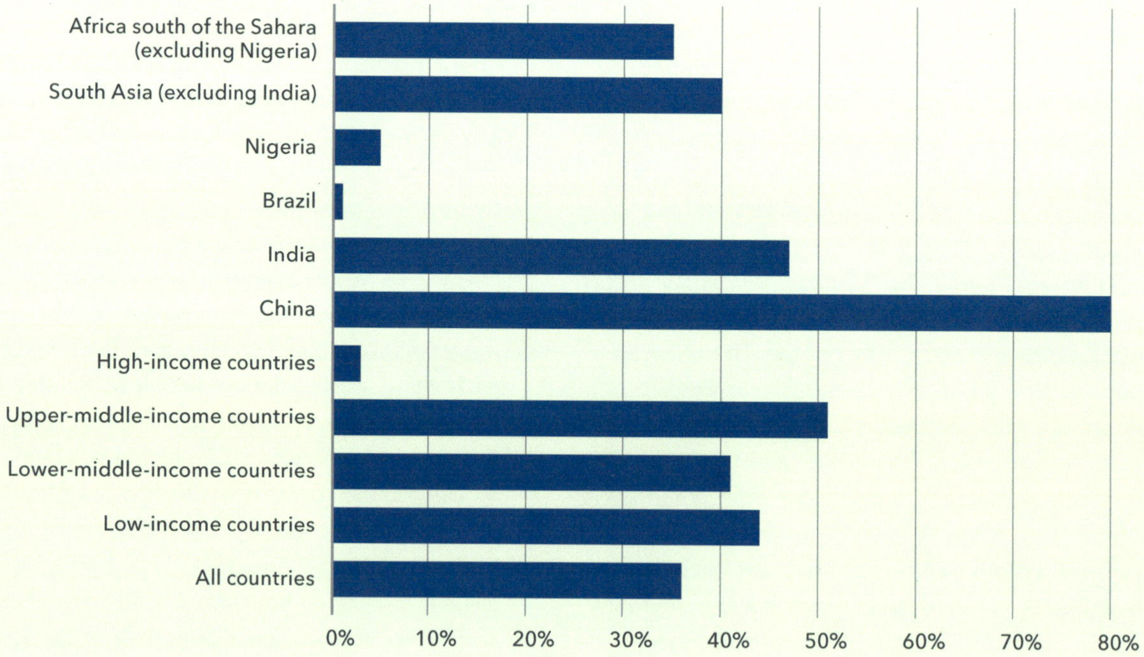

Source: S. K. Lowder, M. V. Sanchez, and R. Bertini, "Farms, Family Farms, Farmland Distribution and Farm Labour: What Do We Know Today?" FAO Agricultural Development Economics Working Paper 19-08 (FAO, Rome, 2019).

and processed foods. Growing demand for these high-value products provides an opportunity for agriculture. But it also presents challenges for millions of small-scale farmers. Expanding and more profitable food markets can encourage the concentration of food value chains in large commercial farms and large-scale processors and distributors (supermarkets), possibly excluding smallholders. To benefit from market opportunities, small-scale producers will have to adjust to ongoing market changes and increasingly stringent food quality and safety requirements in downstream food value-chain segments.

As food systems transform, the emergence of millions of small and medium-sized enterprises (SMEs) in transportation, processing, and distribution—the expanding "hidden middle" of the food supply chain—can promote inclusion of the rural poor. Because food processing, distribution, and services tend to be more labor-intensive, and labor productivity is relatively high in these sectors, food and beverage industries have great potential for creating nonfarm employment. For women in particular, employment in high-value food sector activities has expanded considerably in many countries (see Chapter 4). In Africa and South Asia, midstream activity now represents a substantial portion of agrifood sector GDP, ranging from 25 percent in low-income countries like Rwanda to 60 percent in middle-income countries like Egypt and Indonesia (Box 2). Recent evidence shows that with access to improved infrastructure (roads, storage, electricity, drinking water) and credit, SMEs can thrive and become instrumental in connecting farmers to markets.[7]

To help ensure that food value-chain development is inclusive, efforts to facilitate connections between smallholders, SMEs, and urban markets should be informed by a good understanding of urbanization patterns. About half the total urban population of developing countries, almost 1.5 billion people, lives in cities and towns of 500,000 inhabitants or fewer. Though often ignored by policymakers, geographically concentrated networks of small cities and towns are the places where rural people market their products, buy their seed and other inputs, send their children to school, and access healthcare and other services. These smaller urban centers can play a key role in accelerating the development of rural economies and making them more inclusive.[8]

We propose two sets of policy options to leverage the potential of food systems to boost incomes and create jobs for smallholders and rural workers: (1) promote nonfarm job and income generation through development of the "hidden middle" of agrifood supply chains; and (2) improve farm productivity and incomes by connecting smallholders to markets, with attention to territorial aspects of development.

Box 1 continued

Productive and domestic activities of smallholder farm households tend to be intertwined. Most small farms rely on family labor and produce some food for their own use, but dependence on subsistence farming is becoming less common and participation in food and agricultural markets is increasing (Box 2). Many smallholders supplement low farm-based revenue with income from off-farm work, often in the informal economy. Women make up about 43 percent of the agricultural labor force and some head smallholder households (see Chapter 4).

Smallholders are at higher risk of poverty. Twenty percent of people whose livelihoods are in agriculture are considered extremely poor (living on less than $1.90 a day) and 30 percent are moderately poor (living on less than $3.10 a day). Levels of poverty are notably higher in rural areas—about 18 percent of rural residents are extremely poor, and over 45 percent are either extremely or moderately poor.

Raising farm incomes and improving off-farm options can benefit rural families in terms of nutrition, healthcare, education, and investment in long-term assets. But when smallholders possess little land and human capital and live in isolated communities, they are likely to be poorly integrated into agrifood value chains, with limited access to markets, finance, and services. They are also more vulnerable to weather shocks and input and output price volatility. As a vital part of developing-country food systems, smallholders have much to gain from the potential benefits created by greater inclusion in today's evolving food value chains.

Source: This box draws on FAO, "Smallholders and Family Farmers," Sustainability Pathways factsheet (Rome, 2012); S. K. Lowder et al., "Farms, Family Farms, Farmland Distribution and Farm Labour: What Do We Know Today?" FAO Agricultural Development Economics Working Paper 19-08 (FAO, Rome, 2019); HLPE (High Level Panel of Experts on Food Security and Nutrition), *Investing in Smallholder Agriculture for Food Security* (Rome: FAO, 2013); and FAO, *The State of Food and Agriculture 2015* (Rome: 2015).

BOX 2 CHANGING STRUCTURE OF FOOD SECTORS IN DEVELOPING COUNTRIES

Domestic markets are the primary markets for farmers in Africa and Asia, and their importance is likely to grow. In Africa's food sector, exports make up only 5 to 10 percent of agricultural production and only 10 percent of food consumed is imported. But 80 percent of domestic food supplies in Africa are purchased in markets and handled by private sector value chains, primarily SMEs. Only 20 percent remain within farm households for their own consumption (Figure B2).

Estimates, using the limited data available, find that the share of the agrifood system in GDP ranges between about 30 percent in lower-middle-income countries such as Bangladesh, Egypt, Indonesia, and Viet Nam and 40 to 60 percent in Myanmar and lower-income countries in Africa south of the Sahara (Figure B3). The share of the midstream of the agrifood system is already substantial in all these selected countries (between 19 percent in India and Niger and 57 percent in Egypt) and is growing.

Changes in supply and demand for food products are driving growth in off-farm segments of agrifood systems. First, the share of own-consumption in rural food production has gradually fallen, accompanied by a shift to marketed production of more-profitable vegetables and animal-sourced foods, which has led to a rise in marketing and logistics services. Second, the urban share of the food market has risen rapidly, raising demand for more diverse foods. As cities grow, then, so must supply chains—particularly the transportation and wholesale segments. Third, people are consuming more processed food. As a result, more domestic food processing companies are emerging, and both regional and global companies are entering into national markets. Finally, retailing has evolved over the past several decades, from the marked increase in consumption of food away from home to the rapid spread of fast food chains, restaurants, and supermarkets in Africa and Asia.

FIGURE B2 Supply channels of food consumption in Africa south of the Sahara

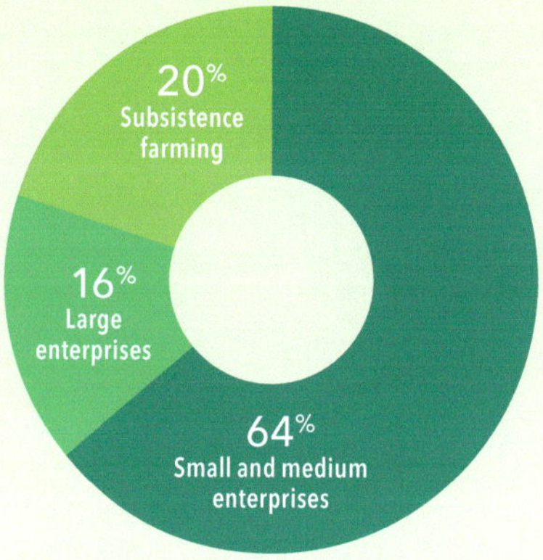

Source: T. Reardon et al., "Overview," in *Africa Agriculture Status Report 2019: The Hidden Middle: A Quiet Revolution in the Private Sector Driving Agricultural Transformation* (Nairobi: Alliance for a Green Revolution in Africa, 2019).

FIGURE B3 Share of agrifood value chain segments in GDP in Africa and Asia

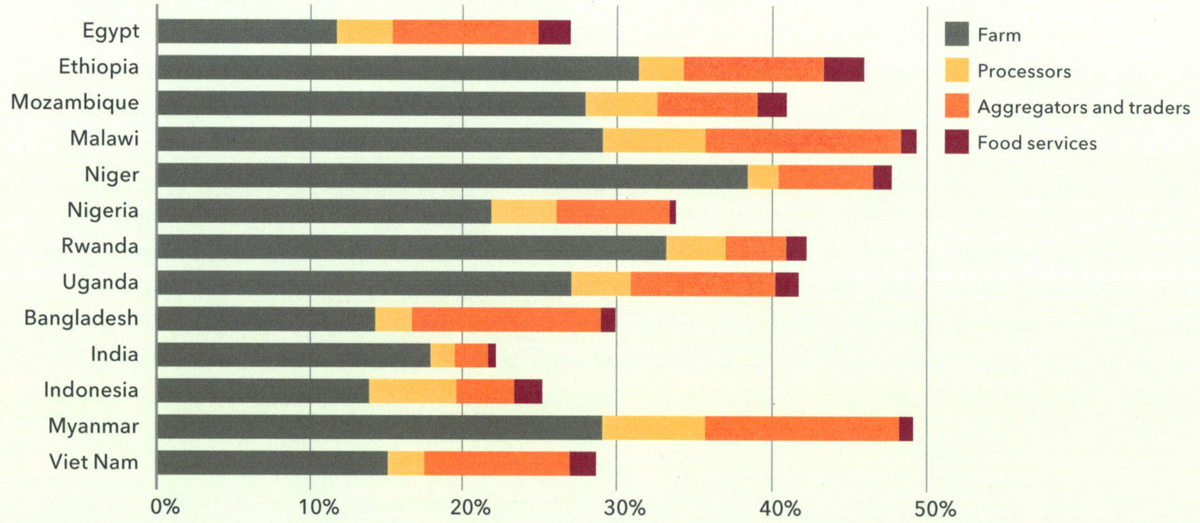

Source: J. Thurlow, "Measuring Agricultural Transformation" (PowerPoint presentation to USAID, Washington, DC, 2020), https://www.slideshare.net/ifpri/aggdp-agemp-measuring-agricultural-transformation.

Source: This box draws on T. Reardon et al., "Overview," in *Africa Agriculture Status Report 2019: The Hidden Middle: A Quiet Revolution in the Private Sector Driving Agricultural Transformation* (Nairobi: AGRA, 2019); T. Reardon and C. P. Timmer, "The Economics of the Food System Revolution," *Annual Review of Resource Economics* 4 (2012): 225–264; C. Barrett et al., "Structural Transformation and Economic Development: Insights from the Agrifood Value Chain Revolution," *Journal of Economic Literature* (forthcoming); J. Thurlow, "Measuring Agricultural Transformation" (PowerPoint presentation to USAID, Washington, DC, 2020),

GENERATING NONFARM AND AGRIFOOD EMPLOYMENT

Policymakers must focus on creating sufficient income and employment opportunities for the developing world's rural population. Since agriculture remains the primary source of food and income for the poor in most low- and middle-income countries, stimulating productivity growth among smallholder farmers is one key to doing so. Development of off-farm activity will also be critical. Nonfarm employment is already more important in rural low-income contexts than often thought. For example, while 70 to 80 percent of rural Africans are engaged in own-farming, recent assessments have shown that it accounts for only a third of their employed time (Figure 1).

In fact, about 25 percent of overall rural employment in both Africa south of the Sahara and lower-income Asia is in the midstream of food supply chains—in areas such as wholesale trade, logistics, processing, and retailing. These agrifood system activities are especially important, particularly in terms of income, for women and youth in peri-urban areas and in areas just beyond. Household survey data for five African countries suggest that income (per full-time equivalent) from nonfarm agrifood system rural enterprises is more than double the income derived from farm activity and also higher than income from non-agrifood system businesses (Figure 2).[9]

Growth of downstream activities—such as packing fruits and vegetables, collecting, refrigerating, and shipping milk, slaughtering animals and preparing and distributing the meat, and collecting and milling feed grains—thus provides opportunities for inclusive economic development. Urban demand for higher-value, more perishable products provides additional income and employment opportunities for actors along food supply chains. Such products tend to have higher economic value because their proper handling requires activities like cold storage and transportation, packaging, and processing that tend to be labor-intensive, both on- and off-farm, when operated through SMEs, possibly even more labor-intensive than the handling of staple foods like grains and pulses.[10] The emergence of these activities creates employment multipliers in rural areas and the small towns that service them.

FIGURE 1 Employment by occupational category in Africa south of the Sahara and Asia

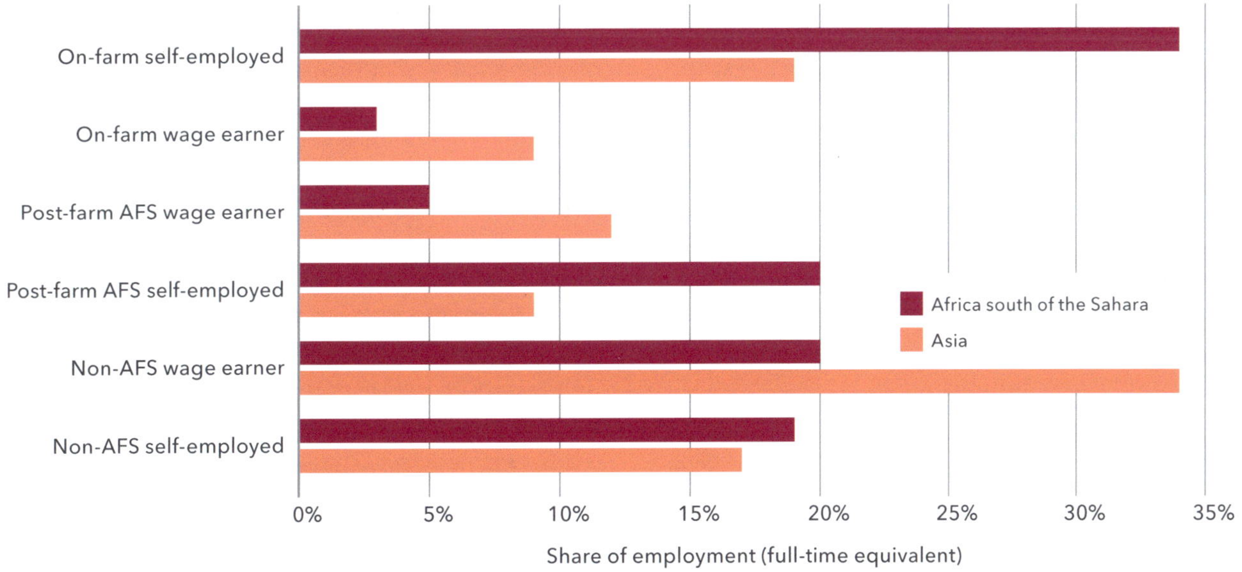

Source: Authors' elaboration based on M. Dolislager et al., "Youth Agrifood System Employment in Developing Countries: A Gender-Differentiated Spatial Approach," IFAD Research Series No. 43 (IFAD, Rome, 2018).

Note: AFS = agrifood system. Employment shares show weighted averages of the values depicted across all countries. The estimates are based on observed data from recent household surveys (2013-2017) in six African countries (Ethiopia, Malawi, Niger, Nigeria, Tanzania, and Uganda) and four lower-income Asian countries (Bangladesh, Cambodia, Indonesia, and Nepal). Dolislager et al. then used these data to estimate a regression model that allowed them to project to regional aggregates.

FIGURE 2 Income from agrifood system and non-agrifood system activities in five African countries

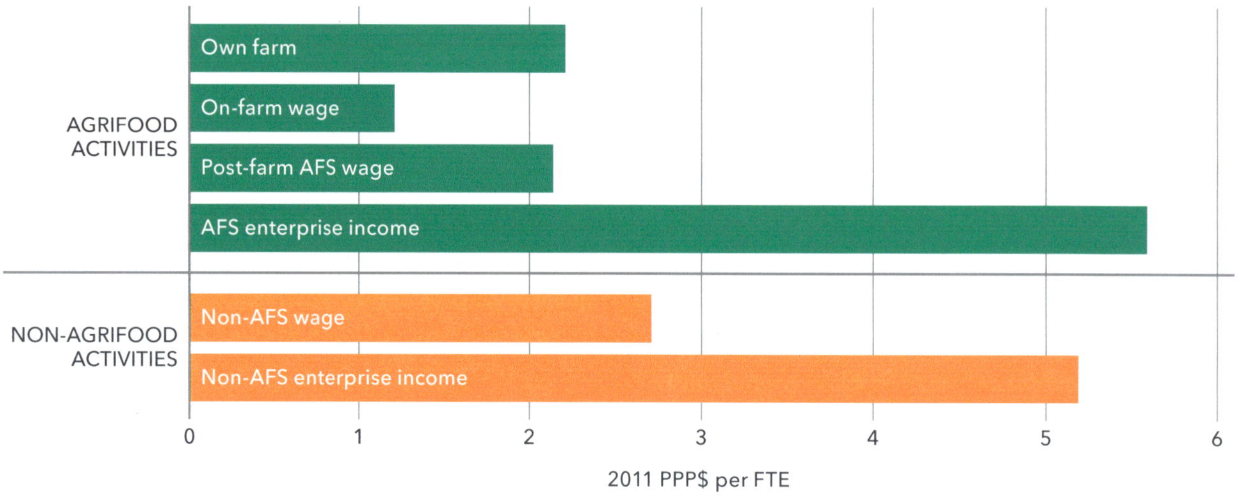

Source: Adapted from M. Dolislager et al., "Youth Agrifood System Employment in Developing Countries: A Gender-Differentiated Spatial Approach," IFAD Research Series No. 43 (IFAD, Rome, 2018).

Note: Based on LSMS-ISA survey data for Ethiopia, Malawi, Niger, Tanzania, and Uganda during 2013-2017. Estimates reflect averages for the five country cases. AFS = agrifood system; PPP = purchasing power parity; FTE = full-time equivalent employment.

Agrifood value chains and other rural-urban linkages are the key to unlocking these opportunities. In low-income countries in Africa and South Asia, rapid expansion of the midstream of food value chains is being driven by the growth and proliferation of SMEs, but has attracted little interest from researchers and policymakers. This "quiet revolution" taking place in food value chains mirrors what happened in other parts of the world in earlier decades.[11] A wide array of formal and informal SMEs dominates this current phase of food system transformation (Box 2). Yet weaknesses remain. Tapping the vast potential of food supply chains to drive inclusive transformation will require public policy support to (1) provide adequate infrastructure, (2) create the right market incentives, and (3) facilitate skills development.

INVESTING IN INFRASTRUCTURE AND MARKET LINKAGES

Adequate rural infrastructure, including quality rural and feeder roads, reliable electricity, and storage facilities, is essential for pro-poor growth and improving rural livelihoods. Inadequate rural infrastructure leaves communities isolated, holds back food value-chain development, contributes to postharvest food losses, and is significantly associated with poverty and poor nutrition.[12]

To stimulate farm productivity and raise farm incomes, infrastructure should be designed to help smallholders access markets. Infrastructure investments should align with support measures that help smallholders overcome other barriers, such as lack of access to credit, improved inputs, or land. For small farmers, such investments help smooth income shocks from seasonality, market volatility, and weather variability. For example, in India, cold storage is reducing the seasonality of the potato supply in Delhi and giving farmers in Agra District new marketing options that counterbalance the power of traditional wholesalers (Box 3).

A comparative analysis of Europe, Brazil, and Chile suggests that infrastructure investment has the biggest impact on market access when it supports a package for connectivity—including improvements in roads, electricity, and communications technology.[13] In Brazil, for example, transport times and costs for individual farmers and drivers have been reduced through infrastructure that provides nodes, such as truck stops, for self-organized transportation of products. In Europe, smallholders in the livestock sector have benefited from infrastructure investments that reduce costs to access local abattoirs, wholesale markets, and Internet ordering systems.

BOX 3 DELHI'S DYNAMIC POTATO VALUE CHAIN

Rapid transformation of India's food supply chains over the past two decades has been marked by expansion of modern retail sales, particularly in urban but also in rural food markets. In the case of the potato value chain, cold storage has given small farmers in Agra District new marketing options. The share of potato farmers using cold storage increased from 40 percent in 2000 to 95 percent in 2009. While traditional storage methods allow for only three months of storage after harvest, new cold storage options mean that potatoes can be stored for seven months. This gives farmers the flexibility to choose among traders competing for their product, including Agra wholesalers, Delhi wholesalers, and rural brokers.

Cold storage operators also provide financial services. Sixty percent of farmers, both large and small, obtain credit using the stored potatoes as collateral. The rapid development of cold storage in Agra District seems to have been driven on the *demand* side by much higher rewards to smallholder farmers when using cold storage and on the *supply* side by the abundant supply of potatoes and rising incomes in Delhi. Government policy has supported this development with subsidies for construction and expansion of cold storage, a reliable power grid, and improvements in roads to Delhi. Because consumers are willing to pay for better-quality stored potatoes, storage costs can be incorporated in prices. Hence, it pays for farmers to store.

Source: Adapted from FAO, *The State of Food and Agriculture 2017: Leveraging Food Systems for Inclusive Rural Transformation* (Rome: 2017).

Public investment in rural infrastructure can also induce forms of inclusive growth that go beyond linking smallholders to markets. For instance, in southern Chile, investment in rural roads and basic services leveraged significant private investment in the salmon aquaculture industry, which reduced poverty by employing rural women in agrifood industries.[14] In central Nicaragua's milk-producing areas, investments in rural roads, cold storage, and milk processing stimulated strong economic and employment growth that benefited traders and large commercial farmers but did not create direct benefits for poor farmers.[15] And in Nepal, investments in roads and bridges moderated food price levels and price volatility.[16]

Investment needs and potential economic synergies are probably best addressed through a territorial or geographic approach.[17] Such approaches include planning of agro-industrial parks, agro-based special zones, incubators, clusters, and agro-corridors, all of which have had varying degrees of success.[18] Infrastructure planning should also support existing "spontaneous clusters" of downstream agrifood businesses, which are too often ignored by national policymakers and donors. Nigeria's thriving maize feed–chicken system provides a good example of a spontaneous cluster driven by large numbers of SMEs in the midstream (Box 4). To further propel agrifood SME dynamics and facilitate deeper integration of smallholders into markets, policies should promote investments that help strengthen the weakest links, which are often the supply of electricity, availability of temperature-controlled storage, and wholesale market development.[19]

Such infrastructure improvements can help dynamize distribution and service networks critical to the development of efficient food supply chains and generate a vast source of off-farm employment. By helping to forge spontaneous SME clusters, infrastructure will further reduce transaction costs for smallholder farmers—directly by connecting them to markets and indirectly by reducing transaction costs for wholesalers (who supply raw inputs to processors). Logistics clusters or hubs such as truck stops tend to emerge near both wholesale markets and SME processors, further reducing the cost of market linkages. This is the case, for example, with clusters of maize milling SMEs in Dar es Salaam and Arusha, Tanzania, located near grain wholesale markets, and likewise first-stage processors and milk collection centers in rural Zambia, some of which are SMEs.[20]

PRICE INCENTIVES AND FOOD QUALITY REGULATION

In addition to infrastructure, adequate price incentives are critical to help small farmers capture a greater share of food system value-added. Price policies that

BOX 4 THE MAIZE FEED-CHICKEN SYSTEM IN NIGERIA

Nigeria has experienced significant economic growth and rapid urbanization in recent decades. Demand for animal-sourced foods has increased, now accounting for about 15 percent of household food budgets in rural areas and 20 percent in urban areas. The maize feed–chicken supply chain has developed as a spontaneous agrifood business cluster in response to the increased demand. In the upstream are roughly 8 million small and medium maize farmers, in the downstream are some 140 million maize purchasers, and intermediating between them are tens of thousands of maize traders, feed and flour mills, and third-party logistics service providers like transporters and warehouse owners.

Feed mills, including both SMEs and large-scale plants, have emerged as a dynamic midstream segment between maize farmers and chicken farmers. As consumption of chicken expanded rapidly during the past 15 years, production shifted from free-range to intensive, feed-based chicken farming–at both large-scale and rapidly growing small- and medium-scale operations. Small-scale fed-chicken farmers have proliferated in the north and south of the country. In response, the feed sector has increased output sixfold over 10 years. Because most of the maize used by the country's feed industry is produced in the north, a long supply chain of maize traders has developed quickly. Much the same has occurred in Nigeria's aquaculture sector, as increased demand for fish has led to expansion of maize-feed-based fish farming.

As a result of the long north–south maize supply chains and the growth of maize demand for both food and animal feed, the maize wholesale sector has also developed rapidly, with attendant growth in transport, warehousing, and handling. This growth has generated substantial employment and enhanced income opportunities of farmers and SME owners and workers in the midstream of Nigeria's maize feed–chicken supply chains.

Source: Adapted from T. Reardon et al., "The Quiet Revolution in Agri-food Distribution (Wholesale, Logistics, Retail) in Sub-Saharan Africa," in *Africa Agriculture Status Report 2019: The Hidden Middle: A Quiet Revolution in the Private Sector Driving Agricultural Transformation* (Nairobi: Alliance for a Green Revolution in Africa, 2019).

help reduce the level and variability of energy costs are especially important. Food processors and distributors rely on consistent, affordable access to electricity. In addition, because much of the equipment used in agrifood businesses in Africa and South Asia is imported, low tariffs facilitate rapid development of food supply chains and job creation.

Helping farmers meet higher food quality standards through regulation and quality certification can also improve market access and incomes for small farmers, making food systems more inclusive. Governments have a responsibility to protect consumers, both SMEs and individuals, from substandard products, whether poor quality seeds and fertilizers or damaged or contaminated food products. Quality certification can also help protect farm-level investments, expand the use of quality seed and fertilizer, increase output, increase SMEs' competitiveness in regional and global markets, and protect consumers. Supermarkets, in particular, which set standards for quality, safety, and consistency, are placing new demands on farmers. For example, food safety concerns become an issue when demand increases for milk, meat, fish, vegetables, edible oils, peanut butter, and similar products, as well as for processed food and food prepared in restaurants (see also Box 3 in Chapter 6).

As large firms take a bigger share of the overall processing sector, SMEs and smallholders will likely face growing challenges in meeting the private sector standards set by supermarkets and large processors. Meeting these standards will require various "threshold investments" in food safety, quality, volume, and consistency by small-scale farmers, which may be cost-prohibitive to asset-poor farmers.[21] In response, governments should consider providing assistance to smaller farms and agrifood operators that lack the means to comply with such initial requirements.

PROMOTING SKILLS AND ENTREPRENEURSHIP TRAINING

Fostering rural entrepreneurship and employment diversification, especially for women and youth, requires the development of general skills, such as those related to running a business, accessing market information, and using information and communications technologies (ICTs). A more skilled labor force in low-income countries would increase agricultural productivity and stimulate the growth of high-productivity services and industrial sectors, and would enjoy access to better-paid jobs. Policies supporting education at all levels are important to inclusive rural transformation, although their impacts are felt only in the long term. Measures to increase the employability of rural youth include strengthening vocational training and basic education, establishing mechanisms for the recognition of informal-sector work experience, and creating greater awareness of job opportunities and labor rights.

However, agrifood businesses in Africa seem to see technical labor skills as less of a constraint on growth than high energy costs and inadequate roads.[22] Further, most firms consider improved basic education and training in social, organizational, and entrepreneurial skills more important than general technical training. In terms of specific technical skills, the most-needed are proficiency in or knowledge of digital technologies, processing techniques, food safety, and ICT-enabled commercial procedures.

CONNECTING SMALLHOLDERS TO MARKETS

The "quiet revolution" in the downstream of food supply chains is also changing farming systems. Growing demand for higher-value food products means that farmers must change the crop production mix. New efficiency requirements and policies have encouraged mechanization and adoption of modern inputs. Often, however, smallholders have been left behind because they lack the resources needed to adapt to the changing food system.

Initiating and sustaining a process of inclusive transformation requires market access and other supports for smallholders to trigger sustainable productivity growth and foster their remunerative participation in food value chains.[23] Here we focus on four instruments for promoting inclusion of smallholders in agrifood supply chains: (1) securing land tenure; (2) promoting inclusive agribusiness models; (3) leveraging the potential of digital technology for smallholders; and (4) enhancing the capacity of farmers and other food chain actors to manage and cope with risks.

LAND TENURE POLICIES FOR INCLUSIVE VALUE CHAIN DEVELOPMENT

Secure land tenure can stimulate agricultural development and improve the well-being of landholders by increasing access to credit and input markets and facilitating land rental and sales markets. Securing land tenure can increase farm productivity, raise the incomes of farmers with limited land, and even facilitate the transition to off-farm activities.[24] Secure land tenure has, for example, been found to improve productivity in Madagascar, provide incentives to farm-level investment in West Africa, and enhance market access in Chad.[25] In many contexts, securing land rights for women in particular can be especially difficult, making it hard for women to access credit and inputs; addressing these issues requires a gender-sensitive approach to the design of land tenure policies and instruments for smallholders (see Chapter 4).[26]

Land tenure plays a role in overcoming hurdles posed by excessive fragmentation of landholdings. An estimated 84 percent of the world's farms are smaller than 2 hectares. In many low- and lower-middle-income countries of Africa and South Asia, average farm size is shrinking, to the point that many farm units using traditional farming practices are no longer economically viable.[27] At the same time, investors are consolidating farmland, and the number of medium-sized farms is increasing in high-potential areas. While the land productivity of small farms tends to be relatively high, the labor productivity of small farms is often low because smallholders lack the necessary scale to access markets or adopt new technologies (underscoring the importance of public rural services and farmers' collective actions, discussed below).[28] Development of efficient land sale and rental markets, which depend on secure property rights, can give farmers access to larger plots that help them achieve economies of scale. Recent evidence suggests that land rental markets are more common than previously thought. For example, in Bangladesh and Togo, 40 percent of holdings are rented or operated under systems other than farmer-owned tenure.[29]

Secure land tenure also supports the development of rental markets for equipment such as tractors and use of improved seeds and other inputs.[30] Agricultural mechanization is critical to boosting productivity because it enhances the performance of other inputs. Mechanization has increased worldwide, especially in those countries that have undergone rapid transformation, and has proved profitable for small-scale farmers. For small farms, equipment rentals and shared use through farmer cooperatives can enable mechanization, as has been the case in parts of East Asia, where use of farm machinery facilitated by rental markets has increased sevenfold since 1985.[31]

Secure land tenure may also increase smallholders' access to water, as it provides incentives to farmers to make long-term investments in both land and water management. However, research on land policies in Ethiopia and Ghana suggests that policies to strengthen land ownership or usage rights on their own may be inadequate and need supplementary support measures.[32]

PROMOTING INCLUSIVE AGRIBUSINESS MODELS

Producer organizations and inclusive forms of contract farming help smallholders overcome constraints to economies of scale and strengthen their access to markets. For instance, producer organizations allow small farmers to engage in collective marketing, which reduces their transaction costs, allows them to share risks, and improves their bargaining power. These organizations link farmers to upstream and downstream actors, thereby helping farmers to obtain better terms, for example, through fairer contract farming schemes.[33] Acting collectively also enables farmers to comply with food quantity, quality, and delivery requirements in supermarket contracts.[34] Small-scale fruit and vegetable producer groups in Kenya, for example, can meet modern market requirements. The country's banana and mango producers benefit from participating in collective marketing schemes. Of Kenya's mango producers, however, medium-scale mango farmers benefit more than small-scale farmers from shared marketing;[35] whereas in China, small-scale farms in general tend to benefit more than medium and large farms, highlighting the importance of context.[36]

Producer organizations also facilitate access to credit, directly by managing microcredit systems and indirectly through innovative arrangements such as warehouse-receipt systems, in which stored produce is used as collateral to obtain short-term loans.[37] Because producer organizations can help farmers meet their financial needs and overcome liquidity constraints, they are especially attractive to smallholders.[38]

Support for small farmers is particularly important today as global input markets consolidate, giving agribusiness input and technology providers little incentive to invest in small farms in developing countries. This context underscores the need for policy interventions that address market failures and respond to small farmers' needs, especially through the provision of public goods such as rural advisory services and support to farmers' collective action.

LEVERAGING THE POTENTIAL OF INFORMATION AND COMMUNICATIONS TECHNOLOGY

Face-to-face extension services and farmers' relationships with buyers are increasingly being complemented—and sometimes replaced—by information channeled through modern ICTs. This is bringing new benefits to smallholders. In India, for example, Internet service provided by a private food conglomerate to rural areas has given farmers access to more information, empowering them in the negotiation of farmgate prices.[39]

Mobile phones in particular are increasing farmers' access to information. Mobile phone coverage and adoption have increased dramatically in developing countries over the past two decades. In Africa, coverage has expanded from less than 10 percent of the population in 1999 to more than 90 percent today. In terms of actual subscribers, 45 percent of Africans now have mobile phone access, and 50 percent are expected to by 2025.[40] In Asia, 66 percent had mobile access in 2019, and 72 percent are expected to by 2025.[41]

Mobile phones effectively shorten the distance between isolated smallholders and other actors involved in processing, transporting, marketing, and regulating farm produce.[42] ICT connectivity allows farmers to seek solutions from peers and expands access to a range of other information sources. For instance, Sri Lanka's FarmerNet, a virtual trading floor, connects traders and farmers via text messaging.[43] Mobile phones have sped up input delivery through e-vouchers and real-time inventory tracking. For example, Nigeria introduced an e-wallet program that delivers seed and fertilizer vouchers directly to farmers' phones. The platform has been extended to deliver other benefits, such as vouchers for nutritional supplements.[44] In Kenya, the Kilimo Salama ("safe agriculture") pilot program uses weather stations to detect excessive and inadequate rainfall and sends a payment to affected farmers through M-Pesa, a mobile money-transfer service.[45] ICTs can also make local access to credit and rural advisory services timelier and more efficient. Finally, it is hoped that ICT-savvy young people in Africa and South Asia will be able to seize new employment opportunities emanating from the widespread deployment of these technologies in agrifood systems.[46]

SOCIAL PROTECTION FOR RISK MANAGEMENT AND LOCAL ECONOMIC DEVELOPMENT

Social protection, for instance in the form of food aid or cash transfers, is crucial to smallholders' risk management during rural transformation and for building resilient rural livelihoods. Social protection allows poor rural households to invest in riskier but more-remunerative livelihood activities. Essentially, these transfers can affect investment decisions via three pathways: (1) managing risks; (2) relaxing liquidity, credit, and savings constraints; and (3) generating spillover effects into the community and local economy.[47]

In a recent positive trend, social protection programs link social transfers to the promotion of rural employment and agricultural production. In Lesotho, a cash transfer program had a larger positive impact on agricultural production when combined with a program to improve homestead gardening.[48] Other programs link public food purchase schemes and school feeding programs to smallholder family farmers as suppliers. A recent study found that a home-grown school feeding program in 10 regions of Ghana had positive impacts when mechanisms were in place to enable the participation of smallholders and ensure access to input support services (Box 5). Impacts tended to be greater for context-appropriate food items—those that are agroecologically suitable and financially viable for small-scale production.[49]

BOX 5 SOCIAL PROTECTION AND INCLUSIVE FOOD VALUE-CHAIN DEVELOPMENT THROUGH SCHOOL FEEDING PROGRAMS

Aulo Gelli (IFPRI)

School feeding programs across the world reach about 368 million children and cost about US$70 billion a year. These programs have been found to improve children's educational levels and cognitive skills, and also to enhance their physical and psychosocial health. Children from disadvantaged families often benefit the most, though effects vary depending on context and the quality of program implementation.

There is also increasing experience linking school meal programs to promotion of farm production. Recently, Brazil and India both have developed large-scale public procurement schemes linked to school feeding programs. Elsewhere, "home-grown" school feeding (HGSF) programs link agricultural development with the market for school meals. HGSF programs typically aim to channel the demand for food for school meals to supply chains that buy from smallholders and small-scale enterprises in transport, distribution, and food preparation. This directly links social protection to efforts to make supply chains inclusive, thus enhancing incomes for small-scale providers and reducing poverty.

Ghana's HGSF program provides school meals through caterers who are directly contracted by the government. Each caterer is responsible for buying food from local farmers and preparing and distributing meals to schools. In a complementary pilot program, caterers have been asked to source food from smallholders and are trained to use local, district-specific menus that meet both dietary preferences and nutritional requirements and recipes that use fresh seasonal foods that can be purchased from producers in the targeted communities. An impact assessment conducted by IFPRI researchers and collaborators indicates that the combined national HGSF and piloted interventions have potential to improve farm sales and incomes in Ghana's poor northern regions. The study also finds significant positive results for educational performance and nutritional status of school-going children.

Source: This box draws on A. Gelli et al., "A School Meals Program Implemented at Scale in Ghana Increases Height-for-Age during Midchildhood in Girls and in Children from Poor Households: A Cluster Randomized Trial," *Journal of Nutrition* (May 2019); and E. Aurino et al., "Food for Thought? Experimental Evidence on the Learning Impacts of a Large-Scale School Feeding Program in Ghana," IFPRI Discussion Paper 1782 (IFPRI, Washington, DC, 2018).

Ethiopia's Productive Safety Net Program (PSNP) is another example with generally positive impacts. The PSNP provides chronically food-insecure rural households with a combination of cash or food transfers (direct support) and transfers through contributions of labor to public works and/or livelihood development. Impact assessments that compared PSNP and non-PSNP households found that both public works and livelihood support programs have had a significant positive impact on participation in nonfarm business activities.[50] On average, the programs increased the probability of engaging in nonfarm activities (mostly in downstream food supply chain activity) by 5 to 7 percentage points. Related assessments found that the program also had positive impacts on crop yields and broader development of the local economy, without creating adverse incentives for agricultural producers.[51]

NO SILVER BULLETS

The potential to create new jobs and better incomes by strengthening food system linkages is enormous, given the growth of food markets propelled by urbanization, income growth, and related changes in dietary patterns. These changes provide opportunities for significant growth in rural incomes and improvements in smallholder livelihoods, as well-integrated networks of downstream activity develop, with new requirements for high-value-added food items, food quality, and food safety. A "quiet revolution" integrating and modernizing food value chains is already underway in Africa and South Asia. Policies can guide this transformation process to ensure that the economic gains from an expanding agrifood sector are shared fairly among supply chain actors, beginning with smallholders, and help address rural needs in regions with the greatest poverty pressures and employment needs.

This chapter offers a range of policy options and instruments that may help make food systems more inclusive for smallholders and rural populations. None of these is a silver bullet, however. Typically, combinations of interventions will be needed to provide the enabling environment for market actors to invest in and innovate for the development of well-integrated food supply chains. To help smallholders connect to markets and help off-farm job creation flourish, policymakers will need to look across the food system to identify and address the weak links.

Such a food systems approach will also be needed to align the objective of developing more inclusive food supply chains with other key food system objectives. We have already underscored the potential trade-off between moving toward an enhanced, consumer-focused food safety regulatory environment and ensuring the financial viability of small farms and food businesses. Other trade-offs concern possible increases in the ecological footprint of longer supply chains and the noncommunicable disease risks associated with the "Westernization" of diets (notably excess intake of salty and sugary processed foods and animal-sourced foods).

To effectively balance such trade-offs, policymakers and analysts will need much better data. At present, policymakers are largely flying blind when it comes to the broader food system. We lack adequate statistics to depict the entire food chain from farm to fork. Farm and household survey data do provide insights into farming systems and farm household income generation, as well as food consumption and nutritional outcomes. But the data from enterprise, market, and field surveys necessary to understand the relative importance and functioning of other parts of the food system, especially those in the midstream, either do not exist or are scattered or buried among different sources. This "hidden middle" extends both to large public assets like domestic wholesale markets and to small private operations, including the millions of food traders, processors, and logistics and service providers.

This data gap hinders policy research and, hence, evidence-based policy guidance. Governments must invest in improved data gathering to provide policymakers with the evidence they need to craft policies that effectively support and promote inclusive value chains across the entire food system.

"Smallholders often struggle to connect with actors in the middle of the food supply chain as a result of limited access to land and inputs and lack of capacity to scale up or implement new practices."

CHAPTER 3

Youth
Including Africa's Young People in Food Systems

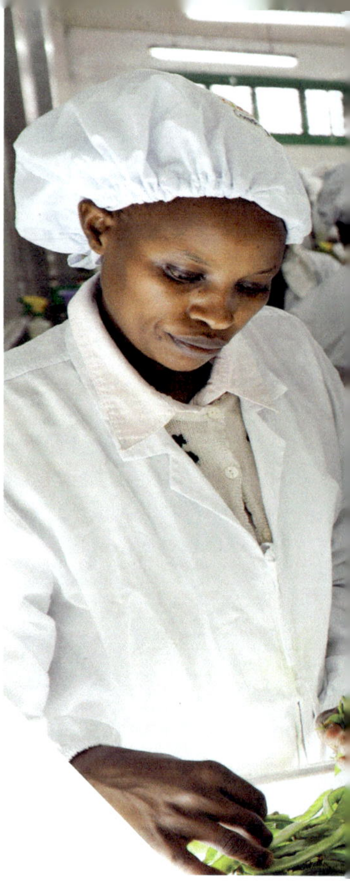

VALERIE MUELLER, GRACIE ROSENBACH, AND JAMES THURLOW

Valerie Mueller is an assistant professor in the School of Politics and Global Studies, Arizona State University, Tempe, USA, and a nonresident fellow in the Development Strategy and Governance Division, International Food Policy Research Institute (IFPRI), Washington, DC, USA. **Gracie Rosenbach** is a research analyst and **James Thurlow** is a senior research fellow in the Development Strategy and Governance Division, IFPRI.

KEY FINDINGS

- Africa's working-age population is growing by 20 million a year, and by 2050 will be growing by 30 million a year, raising questions about whether the region can create enough jobs for young people.
- Concerns should not be overblown—the share of young people in the working-age population peaked in Africa at roughly 38 percent in 2001, not much larger than the peak share in other developing regions during their own "youth bulges."
- Africa's rural areas and food systems will have to play a bigger role in absorbing young job seekers than they did in other regions, given the continued growth of rural populations.
- Opportunities in food systems for youth may be overestimated: Young farmers who are familiar with information technologies are not necessarily more likely than their elders to adopt improved inputs or increase productivity. Nor are they more likely to operate or work for a nonfarm enterprise or to migrate to large urban centers.
- Broad-based development policies that create opportunities for all rural people may do more to support the growing youth population than polices designed specifically for youth.

RECOMMENDATIONS

- Focus on broad-based growth, not just on youth, to create an economic environment in which food system businesses can thrive and generate jobs for both young and old.
- Invest not only in education but also more broadly in sectors such as transportation and energy infrastructure to create inclusive food system opportunities.
- Create more vibrant rural economies and support policies and initiatives aimed at enhancing youth's long-term economic prospects, which in turn will cultivate trust in government among young people.

Almost 20 million people join the working-age population every year in Africa south of the Sahara (henceforth Africa). By 2050, that number will rise to 30 million a year and Africa will become the only region in the world contributing to growth in the global workforce (Figure 1).[1] The absolute scale of Africa's "youth bulge" raises questions about whether, in today's more globalized and competitive world, the region can create enough job opportunities for young people, or whether much of Africa's youth will be "excluded" from the benefits of economic development.

It is not surprising then that many view Africa's rapid population growth with some anxiety: African governments are concerned by the prospect of widespread youth unemployment, which could spark mass protests and threaten stability. Governments elsewhere in the world are concerned by an even greater exodus of African youth from the continent in search of work and a better life abroad.[2] Yet these concerns may be overblown. The challenge of creating jobs for young people is not as daunting from the view of African countries themselves as it is from the perspective of developed countries with smaller populations. In fact, when the share of young people in the working-age population peaked in Africa at roughly 38 percent in 2001, it was not much larger than the peak share had been in other developing regions during their own youth bulges in the 1970s and 1980s.[3] The need to emphasize employment for Africa's youth does not imply that Africa has a "youth problem." Moreover, while Africa as a region is experiencing a youth bulge, its timing varies widely across countries (Figure 2). In South Africa, for example, the share of youth in the workforce peaked in 1976, whereas it will only peak in the Democratic Republic of the Congo in 2027. The pressure to create more jobs for young men and women is therefore unevenly felt within Africa.

What is clear from population and poverty projections, however, is that Africa's rural areas and food systems will need to play a bigger role in providing work for the region's youth than they did elsewhere.[4] A distinguishing feature of Africa is that it continues to experience high rural population growth despite rapid urbanization. Urbanization does create new work opportunities, but the capacity of Africa's cities and towns to absorb enough young job seekers is limited.[5] Thus many (and in some places most) African youth will need to find jobs in agriculture or the rural nonfarm economy.[6]

FIGURE 1 Changes in the global workforce (1950–2100)

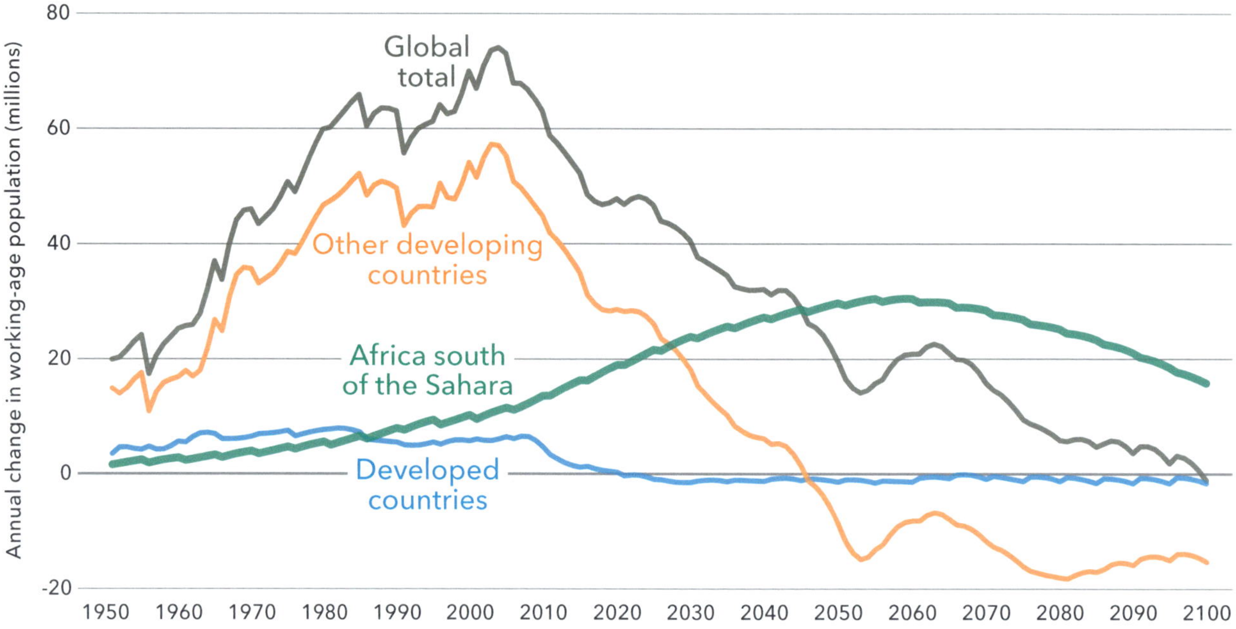

Source: V. Mueller et al., "Africa's Rural Youth in the Global Context," in *Youth and Jobs in Rural Africa: Beyond Stylized Facts*, ed. V. Mueller and J. Thurlow (Oxford: Oxford University Press and IFPRI, 2019).

Note: Uses population data and projections from the United Nations and includes 203 countries/dependencies recognized by the United Nations and the World Bank's country groupings. Working-age population is all people aged 15–64.

FIGURE 2 Peak year and size of African countries' youth bulges (1960–2030)

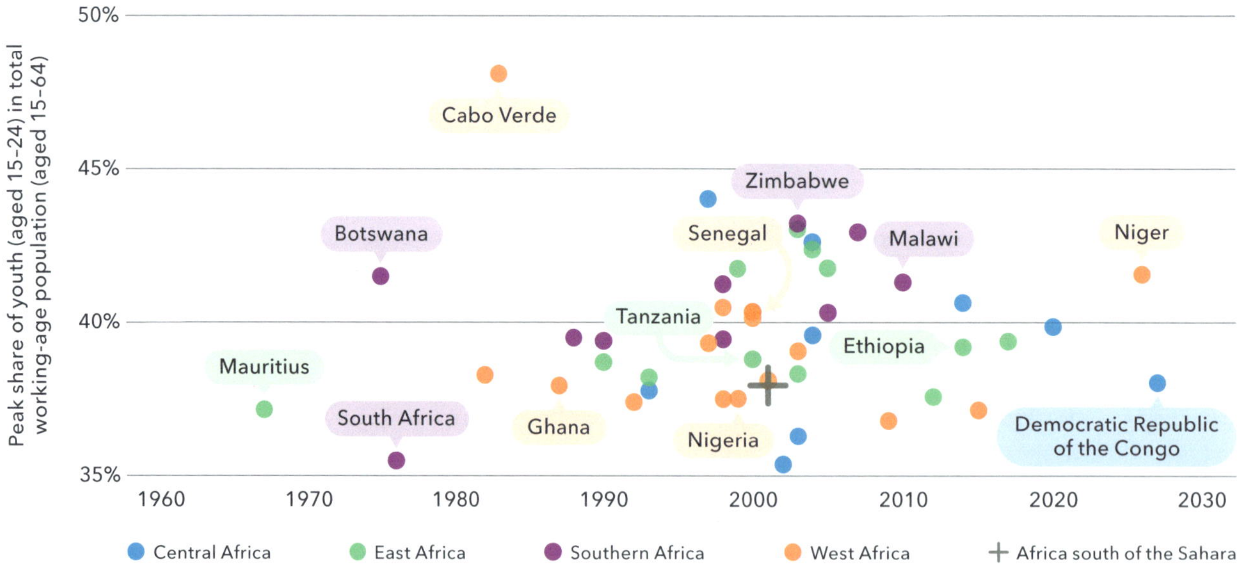

Source: Adapted from V. Mueller et al., "Africa's Rural Youth in the Global Context," in *Youth and Jobs in Rural Africa: Beyond Stylized Facts*, ed. V. Mueller and J. Thurlow (Oxford: Oxford University Press and IFPRI, 2019).

Note: Uses population data and projections from the United Nations and includes 203 countries/dependencies recognized by the United Nations and the World Bank's country groupings.

It is here where concerns about exclusion, or the quality of inclusion, emerge. Increasing land scarcity in many African countries poses a major challenge for would-be young farmers.[7] And young entrepreneurs, like many people in Africa, often lack the financial capital and other resources to start nonfarm enterprises.[8] These problems are well-known and will need to be overcome if rural youth are to take advantage of the opportunities created by growing food demand in African cities.

While Africa's youth bulge stirs anxiety in many observers, others view the continent's growing youth population as cause for optimism. Despite facing significant challenges, these young men and women could play an instrumental role, both on and beyond the farm, in transforming Africa's food systems. Global experience shows that as countries develop, off-farm components of food systems become more important, creating new job opportunities in sectors like food processing and trading (Figure 3). Evidence from East Africa confirms that many of these downstream job opportunities are in rural areas, where up to two-fifths of nonfarm enterprises are in manufacturing, including food processing.[9]

Young people are generally better educated than their elders and are often more comfortable with new technology. African youth may therefore be well-suited, so the thinking goes, to participate in the expanding and more remunerative parts of food systems. By adopting modern farm technologies, starting nonfarm businesses, or migrating to cities and towns for work, young people could help rural households raise and diversify incomes. This more positive outlook on population growth sees young men and women becoming the "agents of change" that Africa's food systems sorely need.

FIGURE 3 Off-farm segments of food systems grow in importance as countries develop (circa 2015)

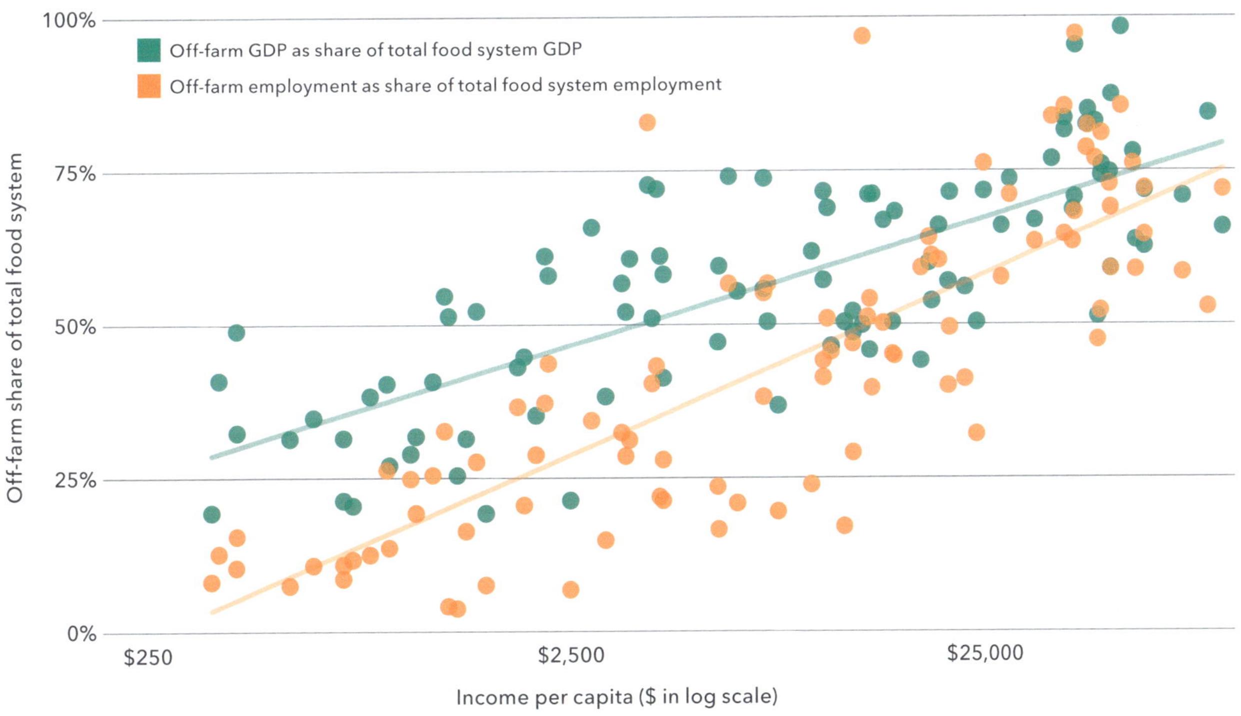

Source: J. Thurlow, "Measuring Agricultural Transformation" (PowerPoint presentation to USAID, Washington, DC, 2020) https://www.slideshare.net/ifpri/aggdp-agemp-measuring-agricultural-transformation.

Note: Uses national accounts and employment data from 96 countries. The downstream agrifood system includes all agriculture-related processing, input production, and trade and transport, as well as accommodation and food services. GDP is gross domestic product and per capita income is based on gross national income, both measured in constant 2011 US dollars.

MYTH VS. REALITY

The new IFPRI book *Youth and Jobs in Rural Africa: Beyond Stylized Facts* cautions against both unbridled optimism about the potential role of rural youth in the food system and undue alarm over the possibility of widespread youth unemployment and consequent political and social instability.[10] The good news is that Africa's agriculture sector is transforming and that fast growth in the downstream components of food systems can be as effective at reducing poverty as growth in agriculture itself (Box 1). This means that investing in food systems should not only create jobs for young people but also help to combat the growing concentration of global poverty in rural Africa.

BOX 1 GROWTH IN DOWNSTREAM FOOD SYSTEMS REDUCES POVERTY

Many studies find that agricultural growth is more effective than nonagricultural growth at reducing poverty. While this is often used to justify public investment in agriculture, it also implies that food systems' historically strong relationship with poverty reduction may weaken as its off-farm components become more important. Simply put, expanding food systems may not create the kinds of jobs that employ the working poor or benefit poor households.

Dorosh and Thurlow (2018), however, find that this apparent trajectory stems largely from the literature's tendancy to overlook the diversity of nonagriculture sectors. Using economywide models for five African countries, the authors estimate poverty-growth elasticities (PGEs) for detailed subsectors. The PGEs in Figure B1 show the percentage decline in the national poverty rate for every one percent increase in national GDP driven by different sectors. Agriculture's PGEs are higher than nonagriculture's overall, but only when nonagricultural subsectors are lumped together. A decomposition reveals that growth in food-system-related nonagricultural subsectors, such as agro-processing and trade and transport, can be as effective as agricultural growth, if not more so, at reducing poverty. These subsectors also grew *much* faster than agriculture in East Africa during 2000–2015, implying that growth in the region's downstream food system may have contributed more to poverty reduction than did growth in agriculture. This bodes well for Africa's rural youth seeking work beyond the farm.

FIGURE B1 Poverty reduction from growth in agrifood subsectors

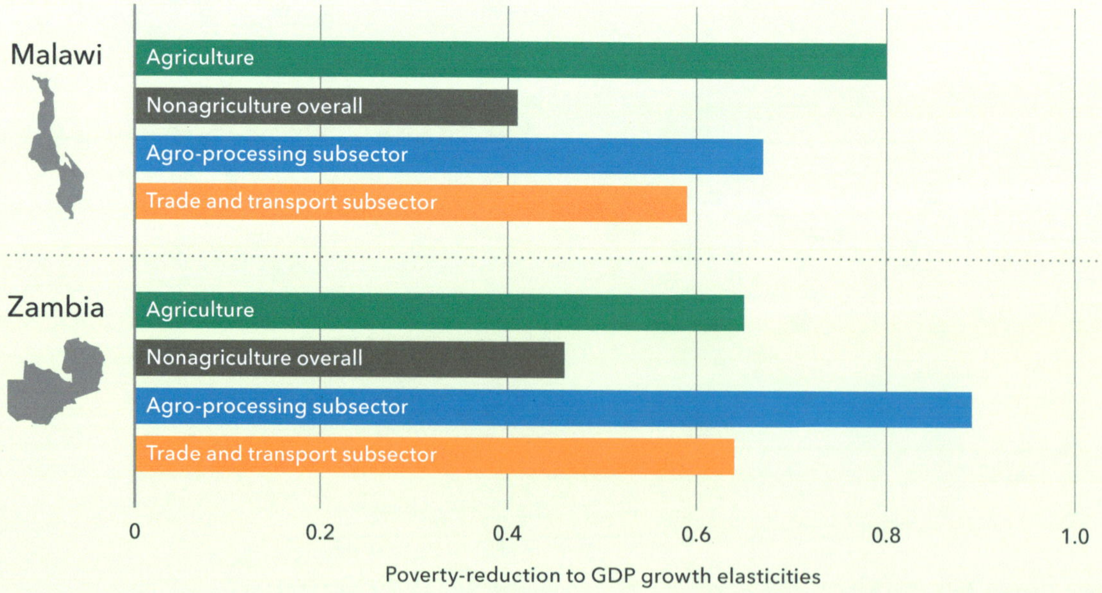

Source: Adapted from P. Dorosh and J. Thurlow, "Beyond Agriculture versus Non-Agriculture: Decomposing Sectoral Growth-Poverty Linkages in Five African Countries," *World Development* 109 (2018): 440-451.

However, the book's five country case studies reveal the diversity of African youth and the dangers of relying on "stylized facts" about how easily young people can participate in food systems. A careful look at the evidence suggests that, contrary to expectations, young people are not always in the vanguard of transformation and are often excluded or are benefiting less than adults are from economic growth. It is therefore critical that we move beyond generalizations and ensure that our understanding of rural youth and their policy needs is grounded in country-level evidence.

It is a common assumption that having more educated young farmers who are familiar with information technologies could lead to higher agricultural productivity, as they would encourage adoption of modern farm inputs and management practices. But evidence that young farmers are more likely to adopt improved inputs is mixed, and even when they do, the effect on productivity may be small. In Ethiopia, for example, youth-headed households are less likely to receive and use advice from extension officers and, more importantly, less likely to use improved technologies such as fertilizers and seeds.[11] Young farmers in Malawi also tend to use fewer modern inputs than older farmers, although this changes once the young farmers become heads of their own households.[12] In Ghana, it is the better-educated farmers, not young farmers per se, who use improved technologies.[13] So it does not automatically follow that having a younger rural population raises farm productivity.

Another stylized fact is that young people, given their higher levels of education, are better-positioned to start and run successful rural businesses, including food processing and food service enterprises. Surveys suggest that this is true in Ethiopia, Ghana, and Tanzania, where youth are far more likely than their elders to operate or work for nonfarm enterprises.[14] Yet even if young people are driving growth in rural nonfarm employment, the nonfarm sector itself is not always a major driver of rural transformation. In Ethiopia, for example, the nonfarm sector remains very small; in Senegal, family-run businesses often result in youth working for free; and in Ghana and Tanzania, young people are more likely to engage in low-productivity occupations, like informal trading, than to work in the formal sector.[15] Even when youth are actively participating in their countries' transformation processes, it may still be their elders who benefit more. In short, young people's education is a crucial asset, but on its own can neither guarantee individual success nor singlehandedly ensure rural transformation. Of course, investing in education, especially for girls, tends to reduce fertility rates, and this would eventually slow down population growth and reduce the number of young job seekers.[16]

Finally, it is widely believed that rural youth are more likely than their older counterparts to migrate. Evidence from Malawi and Tanzania supports this. Moreover, when young people migrate to cities and towns, their higher-paying urban jobs help to diversify rural households' incomes through remittances.[17] Yet few young Malawians or Tanzanians actually move to urban centers. Rather, they are far more likely to migrate from one rural area to another, and doing so imposes a cost on those left behind (Box 2). Even when youth in Malawi and Tanzania do migrate to cities and towns, many continue to devote at least some of their time to working in agriculture. In fact, evidence from Ghana suggests that jobs associated with the food system are more prevalent in peri-urban areas surrounding small towns than in big cities. So, while youth are more likely than their elders to migrate, the importance of urban migration, especially to big cities, for young people in finding jobs beyond the farm should not be overstated.

In summary, young men and women are already participating in Africa's food systems, and in some countries, they are helping to raise farm productivity, start nonfarm businesses, and take advantage of urban job opportunities. However, their continued inclusion is not assured without supportive policies and public investment programs.

BETTER POLICIES, NOT YOUTH POLICIES

Creating rural employment in the food system will be key to ensuring that African economies can avoid the prospect of widespread youth unemployment. However, Africa does not have a "youth problem," but rather faces the broader challenge of promoting inclusive economic transformation. Africa must provide better jobs for younger and older workers alike, all of whom aspire to higher living standards and better working conditions.

BOX 2 YOUTH MIGRATION HAS A COST FOR THOSE LEFT BEHIND ON THE FARM

Although rural-to-urban migration among youth (ages 15–24) in Africa is quite low, young people are increasingly moving *within* rural areas for work. Given the high prevalence of youth migration, Mueller et al. (2018) shed light on the net benefits that youth migration yields for rural households in Ethiopia and Malawi. While the migration of the sons and daughters of household heads can increase the incomes of typical agrarian households, it may also leave them with a shortage of labor to prepare, cultivate, and harvest the land.

Using panel data from the Living Standards Measurement Study–Integrated Surveys on Agriculture in both countries, the authors find that the migration of young men, in particular, increases the burden of farming activities on family members left behind. In Ethiopia, adult female family members experience increasing time burdens on the farm as a result, while in Malawi it is younger male family members who must pick up the slack. In spite of the induced labor constraint on households, the migration of young men results in an overall net income gain to households in Ethiopia. In contrast, migrant households in Malawi face net losses. These losses coincide with an increase in household expenditure on hired labor, presumably to substitute for the absent migrant on the farm. The findings highlight the increasing importance of diffusing labor-saving technologies in these areas in order to enable households to diversify their income outside of the agriculture sector without compromising their main livelihoods or the welfare of other family members.

Source: Adapted from V. Mueller, C. Doss, and A. Quisumbing, "Youth Migration and Labour Constraints in African Agrarian Households," *Journal of Development Studies* 54, no. 5 (2018): 875–894.

Older workers in Africa will also need to move out of agriculture and into the broader food system and economy as the process of structural change unfolds in their countries. That said, even a broad-based development strategy that targets all workers must pay special attention to its effects on youth employment.

Creating jobs for young men and women is already a major policy goal in most African countries, but the means of achieving this goal are not always well-represented in current policies.[18] Experts in the development community are largely divided between those who promote youth-specific policies and those who advocate broad-based development policies. The former tend to be strongest on labor supply issues, such as promoting self-employment and improving youth education, but they are weaker on labor demand issues, such as stimulating private sector job creation. This uneven focus implies that it is an underinvestment in young people's capabilities that is preventing them from finding decent work.

However, as discussed above, education alone is not a panacea. Broader-based policies and investments are needed to create an economic environment in which businesses that employ and are run by young and older people alike can thrive. Young people need policies that produce better economic opportunities—not policies that narrowly focus on youth capabilities. This means investing not only in education but also in areas such as transportation and energy infrastructure.

While the scale of policy reforms and actions needed to create more and better jobs for Africa's growing workforce is daunting, there is now a clear alignment of interests and incentives. African governments have made youth employment a policy priority, and African youth are demanding policies that improve their job prospects. While evidence suggests that young people in Africa are only slightly more likely than adults to protest, youth protests are more often motivated by concerns about unemployment.[19] To avoid the possibility of dissatisfied young people taking to the streets *en masse*, governments need to cultivate greater trust among the youth, not only by creating more vibrant rural economies, but also by improving dialogue with young people to enhance their role in decision-making processes.[20] Young women and men need assurances that government policies and initiatives are aimed at enhancing the youth's long-term economic prospects rather than simply at mobilizing short-term political support. African governments can prevent widespread youth unemployment just as other developing regions have, but doing so will depend on policies that help whole economies flourish and create better jobs for everyone.

"African governments have made youth employment a policy priority, and African youth are demanding policies that improve their job prospects."

CHAPTER 4

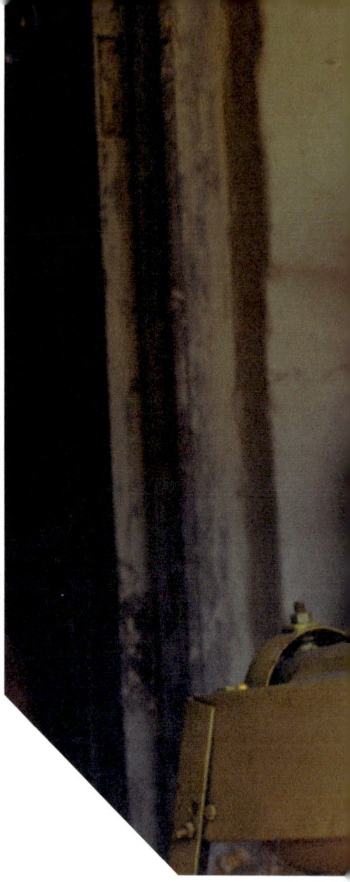

Women
Transforming Food Systems for Empowerment and Equity

HAZEL MALAPIT, RUTH MEINZEN-DICK, AGNES QUISUMBING, AND LAURA ZSELECZKY

Hazel Malapit is a senior research coordinator and **Agnes Quisumbing** is a senior research fellow in the Poverty, Health, and Nutrition Division, International Food Policy Research Institute (IFPRI), Washington, DC, USA. **Ruth Meinzen-Dick** is a senior research fellow in the Environment and Production Technology Division, IFPRI. **Laura Zseleczky** is a program manager in the Director General's Office, IFPRI.

KEY FINDINGS

- Women are actively involved in food systems in many roles, but their contributions are often not formally recognized, and they face obstacles to engaging on equitable and fair terms.
- Together with changing diets, transformation of food systems toward more efficient and sustainable production processes and longer value chains offers new opportunities and challenges for women's participation.
- Transforming food systems for inclusion means not just ensuring women's participation and access to benefits but also their empowerment to make strategic life choices.
- Entrepreneurship is often touted as a key to empowering women, but evidence indicates that it may not empower women if limited to small, household-based enterprises.

RECOMMENDATIONS

- Increase women's decision-making power and control over resources and assets (such as credit, land, training, transport, and technology) within households and communities.
- Raise women's voices in key processes such as negotiations with market actors, research decisions, and political processes.
- Include women and consider women's needs and preferences in the design of institutions, including property rights, financial institutions, and access to information and education.
- Collect and evaluate more data relevant to women's empowerment within food systems, including on capacities, motivations, and roles in value chains.
- Encourage private sector initiatives to foster women's empowerment, including adoption of standards for gender equity, women's empowerment, and women's leadership.
- Ensure that food system transformations do not disempower women by increasing workloads or reducing decision-making power, but rather create a virtuous cycle of inclusion and empowerment to benefit women and men.

Women are actively involved in food systems in a range of roles from production and processing to retailing and consumption. Women grow and manage crops, tend livestock, work in agribusinesses and food retailing, prepare food for their families, and much more.[1] But women's contributions to food systems are often not formally recognized, and women often face constraints that prevent them from engaging on terms that are equitable and fair. In many countries, women have less schooling than men, control fewer resources, have less decision-making power over household income, and face time constraints because of their triple burden of productive, domestic, and community responsibilities.[2]

Gender also intersects with other spheres of vulnerability and identity—including ethnicity, age, and poverty—to further impact how women engage in food systems. For instance, across the food system, young women seeking to become entrepreneurs can face multiple constraints based on gender, age, and the nature of work in the informal sector. If they are married, they may face additional challenges, falling through programming cracks if they are no longer in school, have to care for young children or other family members, or lack the resources required to do business in the food system.[3] The transformation of food systems toward more efficient and sustainable production processes and longer value chains, in combination with shifts in diets toward greater consumption of prepared foods, offers a range of new opportunities for women, but may also create new barriers to participation.

Transforming food systems to be more inclusive will require approaches that not only enable women to participate and benefit equally but also empower women. In this context, "empowerment" is understood as a process by which people expand their ability to make strategic life choices where they were previously denied that ability.[4] The reach-benefit-empower framework[5]—developed to distinguish between agricultural development project approaches that *reach* women as participants, those that *benefit* women, and those that contribute to *empowering* women—can be a useful lens to explore how food systems can be transformed to be more inclusive and gender-equitable. Reaching women as participants does not ensure that they will benefit from a project, and if they do accrue benefits such as increased income or better nutrition, that does not ensure that they will be empowered to

BOX 1 WOMEN'S EMPOWERMENT ACROSS VALUE CHAINS IN BANGLADESH AND THE PHILIPPINES

IFPRI piloted the Women's Empowerment in Agriculture Index for Value Chains (WEAI4VC) in Bangladesh and the Philippines, two Asian countries with very different sociocultural contexts. The Philippine study looked at empowerment across different value chains, whereas the Bangladesh study examined whether empowerment differs depending on the role that women play in the value chain. Figure B1 compares the extent of disempowerment for men and women (longer bars indicate greater disempowerment) and the relative contributions of different indicators to disempowerment. The WEAI4VC pilot studies show how researchers and practitioners can measure, compare, and identify sources of disempowerment for women and men across value chains and find ways to address them that are targeted specifically to each value chain or role within a value chain.

DOES PARTICIPATION IN VALUE CHAINS EMPOWER WOMEN?

Investigation across four Philippine value chains indicates that participation alone cannot ensure increased empowerment. Gendered stereotypes contribute to disparate workloads and inhibit women from seeking jobs in other parts of the chain. For example, women are often engaged in tying, planting, and drying seaweed and earn lower incomes based on piece rates, while men do the more strenuous work of diving to attach seaweed lines to stakes and earn a higher daily rate. Gender norms related to mobility and heavy labor also intersect with women's more limited access to capital and knowledge of specialized markets, making it more difficult for individual women than men to expand a trading business in the coconut and seaweed chains.

In Bangladesh, women participate in value chains by providing household labor, for which they are not individually compensated. In contrast to the Philippines, Bangladeshi women do not typically maintain control over the income generated by their work, and their heavy workload is a major contributor to making them less empowered. Overall levels of empowerment were found to be lower for both men and women in Bangladesh compared with the Philippines. Women in Bangladesh are less empowered than men, and women's empowerment varies greatly with livelihood activity, while men's empowerment is relatively similar across livelihood activities.

ARE SOME VALUE CHAIN ROLES—PRODUCERS, ENTREPRENEURS, OR WAGE WORKERS—MORE EMPOWERING THAN OTHERS?

In the Bangladesh study, women in producer households were found to be more empowered than those in entrepreneur or wage-work households. Because working away from home is less socially desirable, women entrepreneurs and wage workers may be more susceptible to losing the respect of community members. Compared to women in producer households, women in entrepreneur and wage-work households have little say in productive decisions.

DO VALUE CHAINS FOR PARTICULAR COMMODITIES OFFER BETTER OPPORTUNITIES TO EMPOWER WOMEN?

In the Philippines study, women in the abaca and coconut value chains are less empowered relative to those working in swine and seaweed, but the specific areas of disempowerment vary from chain to chain (Figure B1). However, some of the same gender issues exist across value chains, highlighting the need for transformative approaches that can address structural social and gender norms, such as promoting gender awareness in communities and schools, targeting not only women and girls but also men and boys.

Source: A. Ahmed et al., *Tracking Empowerment along the Value Chain: Testing a Modified WEAI in the Feed the Future Zone of Influence in Bangladesh* (Dhaka, Bangladesh, and Washington, DC: IFPRI, 2018); H. Malapit et al., "Empowerment in Agricultural Value Chains: Mixed Methods Evidence from the Philippines," IFPRI Discussion Paper 1881 (IFPRI, Washington, DC, 2019).

Box 1 continued

FIGURE B1 Women's and men's disempowerment along value chains in the Philippines and Bangladesh

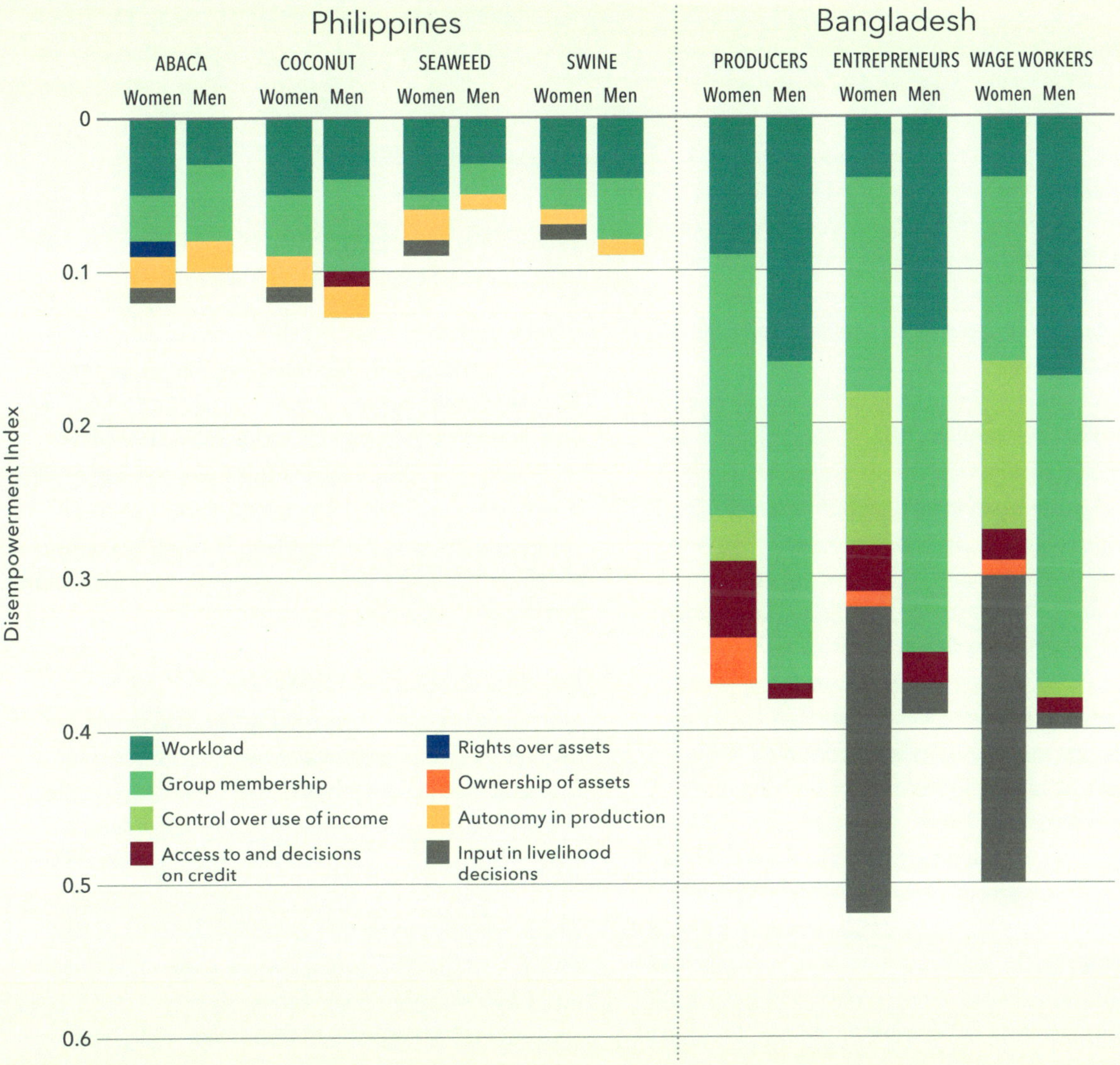

Source: Women's Empowerment in Agriculture Index for Value Chains quantitative surveys, IFPRI.

Note: Autonomy in production and rights over assets were included only in the Philippines study.

control that income or choose foods for their households.[6] Measuring the effect of a program's ability to reach, benefit, or empower women will require indicators specific to each approach. For example, reach can be measured by tracking the number of women who participated in a food system project, and benefit can be measured by assessing women's nutritional outcomes. Women's empowerment encompasses many dimensions that can be measured by indicators such as sole or joint decision-making over productive activities like farming or fishing; ownership, access, and decision-making power over productive resources such as land or credit; sole or joint control over income and expenditures; membership in economic or social groups; and allocation of time to productive and domestic tasks, and satisfaction with the time available for leisure activities.[7]

Studies examining gender dynamics and women's empowerment along value chains in food systems or within specific sectors illustrate that many interventions reach women and even benefit women, but there are still barriers to women's empowerment. A review of interventions promoting high-value agriculture in Bangladesh, Burkina Faso, Mozambique, and Uganda found that involving women in the projects helped increase production, income, and household assets. Project benefits, however, were constrained by existing gender norms in asset use and control—in most cases men's incomes increased more than women's and the projects did not reduce the gender asset gap.[8] Entrepreneurship is often suggested as a key to empowering poor rural women.[9] However, evidence from Bangladesh and the Philippines indicates that entrepreneurship may not be empowering for women if limited to small, household-based enterprises, which typically are not very lucrative and can add to women's workload. The benefits of entrepreneurship may only materialize as businesses grow and owners can start hiring other workers and retain more of the profits (Box 1).[10]

Ensuring that women's contributions to food systems are recognized—by their families, communities, policymakers, and society more broadly—and that women can make strategic choices about their involvement in food systems has benefits for all of society. Women's empowerment can improve agricultural productivity, household food security and dietary quality, and maternal and child nutrition.[11] Women's activities throughout food systems range from growing food and generating income through agricultural production, marketing, and retailing to helping ensure adequate nutrition through decisions about food purchases and preparation to demanding better food environments through political participation. Given the vital role that women play in food systems for themselves and their families, it is imperative that they can engage equitably and that constraints on their empowerment be addressed through changes to policy, programming, and norms.

CREATING INCLUSIVE AND EMPOWERING FOOD SYSTEMS FOR WOMEN

What would a food system that includes and empowers women look like? Among other things, women's roles and contributions would be recognized and valued, women would be able to exercise control over resources and assets, they would have a voice in key processes, and institutions would be supportive of women. Work is underway in many countries and communities to transform food systems toward this vision, but there is still a long way to go.

INCREASE WOMEN'S DECISION-MAKING POWER AND CONTROL OVER RESOURCES AND ASSETS

Increasing women's decision-making power and control over assets within their households and communities is a key step toward inclusive food systems. A recent systematic review finds that their access to assets like credit, land, training, and transportation is limited, which reduces women's choices and influences their ability to engage in more lucrative, larger-scale activities. Beyond production, higher-paying jobs that require specialized training tend to favor men, while women are constrained by lack of time (due to their domestic responsibilities), limited transportation options (for example, due to taboos around riding bicycles and motorcycles), and lack of sanitation facilities in markets and other public spaces.[12] In some areas, men have moved out of agriculture or migrated away from rural areas, leading to a "feminization" of agriculture (Box 2). This can present opportunities for women to gain greater decision-making power and higher earnings in key parts of food systems, but it can also increase women's workload and financial burdens.[13]

BOX 2 FEMINIZATION OF AGRICULTURE

Many rural landscapes are rapidly transforming, driven by a wide range of factors that are shifting labor patterns and decision-making in agriculture. Some of these key drivers include commercialization of value chains, climate change, technology, and migration.

HOW ARE THESE CHANGES AFFECTING WOMEN, PARTICULARLY IN PLACES WHERE MEN ARE MOVING OUT OF AGRICULTURE?

The extensive literature on "feminization of agriculture" describes two distinct views of how these transformations are affecting women's workloads and agency, including decision-making authority and control over resources. On the one hand, these changes can create new opportunities for women to engage in paid employment and commercial agriculture. These opportunities can increase women's incomes and their visibility and voice in their households and communities, stimulating a virtuous cycle of economic empowerment. On the other hand, women may carry additional burdens of labor and responsibility without the agency and resources to take full advantage of new opportunities. While both views acknowledge that these processes affect women and men differently, it remains unclear how gendered patterns of agricultural labor are changing and what the net impacts of these changes are on women and men in terms of work and agency.

While the global evidence confirms that in many places women's labor force participation in the agriculture sector is increasing, it is often concentrated in seasonal, casual, or unpaid work. Women are often paid less and face poorer working conditions compared with men doing similar jobs. Women's overall work burdens increase substantially as they avail of new opportunities because they often remain responsible for productive and domestic work in the home. However, women's increased involvement in paid work and engagement in agricultural decision-making have the potential to close gender gaps, particularly when women take on management or supervisory roles, access labor-saving technologies and training, and gain greater access to and control over resources and incomes.

HOW CAN AGRICULTURAL RESEARCH AND POLICIES ENSURE THAT WOMEN AND MEN BENEFIT EQUALLY FROM THE INNOVATIONS AND TRENDS THAT ARE AFFECTING RURAL LABOR MARKETS?

Interventions and policies must be designed based on a clear understanding of the context and of how both labor patterns and gender relations are changing, often in response to new opportunities or setbacks. Changing labor patterns also have distinct impacts on different groups of women and men based on age, caste, ethnicity, or other socioeconomic characteristics. Understanding these complex trends can lead to more nuanced, and therefore more appropriate, policy prescriptions. Critical to this understanding are more detailed and high-quality quantitative and qualitative data, which we need in order to better understand the intrahousehold dynamics that underlie livelihood decisions—for example, the drivers or impediments to joint decision-making, the nature of asymmetric information between spouses, and women's and men's use of time. Finally, detailed data on the policies, institutions, and norms in a given setting can provide insights into the potential of agrarian change for bringing about gender equality.

Source: Adapted from C. Doss et al., "From Feminization of Agriculture to Gender Equality," in *Agricultural Research Progress Towards Gender Equality: Taking Stock and Setting a Forward-Looking CGIAR Gender Research Agenda* (working title), ed. R. Pyburn et al., CGIAR Collaborative Platform for Gender Research (forthcoming, 2020).

Women's control over assets and decision-making power in food systems can be supported by enhancing their negotiating power vis-à-vis market actors through fair contracting or payment schemes. In Uganda, the Farm and Family Balance project is working with the country's largest sugarcane processing company, Kakira Sugar Limited, to increase women's involvement in sugarcane marketing and management activities by encouraging the transfer or registration of a contract for a sugarcane block—the parcel of land on which the sugarcane is grown—from husbands to their wives. The contract entitles the wife to receive inputs on credit, cash advances, and the final payment associated with the block. Initial findings indicate that overall acceptance of the intervention was high (70 percent), suggesting that simple encouragement can effectively nudge men to include their wives in household commercial activities.[14] This also highlights the importance of working with men as well as women in programs that may shift or alter traditional gender roles and norms.

RAISE WOMEN'S VOICES IN KEY PROCESSES

In addition to ensuring that women's engagement in activities throughout the food system is equitable and fair, it is critical that women's voices be heard in processes related to food systems, such as research, and in the contexts in which food systems are embedded, such as political processes. For example, agricultural research for development is an essential pathway for food system transformation. It contributes to improved management practices, production, processing, transportation, and more, and women's priorities and preferences—such as for food crops with certain nutritional or taste qualities or that do not require excessive labor—must be part of the research process. In Kenya, GROOTS—a national movement of grassroots women-led community-based organizations—is working with the agricultural extension system to provide input on the types of climate-smart technologies preferred by the members.[15] Recognizing women's needs and priorities in the early stages of research is an important step toward ensuring that women benefit from the results.

Political mobilization is also a central avenue through which women's voices can influence the policies that shape the food system. By voting or by becoming policymakers themselves, voicing support or dissent for key policies, and participating in other civic processes, women can affect the way food systems operate. Evidence from India shows that women who belong to women's self-help groups are more politically engaged and make use of a greater number of public entitlement schemes.[16] Membership organizations, such as the Self Employed Women's Association in India (Box 3), also offer opportunities for women to work together to address the specific challenges they face in particular segments of the food system.

ENHANCE INSTITUTIONAL SUPPORT FOR WOMEN

Ensuring that institutions are supportive of women is another critical factor for inclusive food systems. For example, formal laws and informal systems governing property rights impact women's abilities to invest in their land or businesses, access credit, and diversify their livelihoods. There is strong evidence that women's land rights affect the extent to which they make decisions about household consumption, human capital investment, and transferring resources to the next generation, though further research is needed to investigate these links more directly.[17]

Financial institutions also hold great potential for empowering women, but when financial systems are not designed with women's needs and preferences in mind, they risk exacerbating gender wealth gaps. Financial inclusion requires a gender-transformative approach that focuses explicitly on expanding women's opportunities and decision-making power; strengthening relationships and improving negotiation dynamics at multiple levels (home, workplace, market, and within financial institutions); and promoting broader enabling policies, regulatory frameworks, and sociocultural norms. In practice, this might mean including gender analysis in financial product design and service delivery, conducting "gender-smart" due diligence to better understand the contexts in which clients are operating, or evaluating performance against gender empowerment outcomes. For example, a partnership between CARE International, PostBank, and local partners in Uganda is introducing a mobile banking product specifically designed for women's priorities (such as saving for school fees or healthcare), providing financial counseling sessions,

BOX 3 THE SELF EMPLOYED WOMEN'S ASSOCIATION: EMPOWERING WOMEN IN INDIAN FOOD SYSTEMS

The Self Employed Women's Association (SEWA) has members who work across the entire food system in India, from small-scale producers to processors, retailers, prepared food vendors, and consumers. As a membership organization, SEWA is aware of the challenges faced by each group and seeks to respond to its members' priorities and preferences to enable them to make strategic choices related to their livelihoods. There are some common issues, such as lack of credit or insurance, which SEWA addresses through financial services, as well as aspects unique to each group. For producers, information about technologies to adapt to climate change is a priority. For processors, training on food safety standards and changing consumer preferences is key.

When SEWA noted that many of its rural members did not have access to safe and affordable food supplies, it started its own brand of products, RUDI (Rural Distribution Network), that includes flour, spices, and other staples. RUDI is a for-profit agribusiness company that connects the farmer to end users. It is fully owned and operated by over 250,000 small-scale women farmers. The company has its own procurement channels, processing centers, packaging units, and distribution network. The smallholder farmers sell their produce to RUDI. The produce is graded, processed, and packaged into affordable small packages and redistributed in the villages by SEWA's salesforce–called Rudibens or Rudi Sisters. RUDI brings nutrition and food security to over a million households today. Women from various sectors of the food system are actively involved in, benefit from, and control this process–from the farmers who receive fair returns to the landless laborers who are employed as sales people.

Source: Adapted from N. Benni and R. Barkataky, "The Role of the Self Employed Women's Association (SEWA) in Providing Financial Services to Rural Women" (FAO, Rome, 2018); and SEWA, "Rudi Products and Rudi Multi Trading Co Ltd," accessed December 4, 2019, www.sewa.org.

and integrating a research component to better understand factors affecting uptake of the program and users' experiences.[18]

Access to timely information is another critical factor to ensure women's equitable inclusion in food systems. Women's access to information can be facilitated or inhibited by technologies, programs, or institutions. For instance, in Tanzania, under a rice warehousing system that gave information only to the men who delivered the crop, women lost out on information about what rice stocks the household had for domestic use or sale to meet household expenses. As one woman explained, "as days go by, you can't go daily to check them [the sacks], since you aren't the one who signed for it inside there."[19] Mobile phones can facilitate access to banking and government programs. They can also connect producers with information such as extension services or real-time updates on market prices and weather–information that is increasingly important in the context of climate change.[20] Yet women own and use mobile phones at lower rates than men, and thus may have less access than men to services and information provided via mobile phones.

Ensuring that girls and women have equal access to educational and training opportunities and institutions is also essential for their inclusion and empowerment in food systems.[21] Closing the gender gap in basic education can help the next generation of women break out of poverty and make informed and strategic choices about their livelihoods, the food they and their families consume, and the types of policies they demand from political leaders. Moreover, vocational training can ensure that women have the skills required for off-farm and entrepreneurial opportunities throughout the food system. In a virtuous cycle, inclusion in the food system through well-paying jobs can help women and their families make further gains, including investment in the next generation's education and skills.

TAKING ACTION AND MOVING FORWARD

Food systems are transforming in many ways, and as the world faces demographic shifts and global challenges such as climate change, it will be increasingly urgent to ensure that changes open opportunities within food systems without putting additional burdens on women.

MORE DATA. An essential first step toward more gender-equitable food systems is to better understand where there are opportunities for women's empowerment, particularly beyond the well-studied areas of production and processing. A strong body of research looks at women's roles in market-oriented agriculture and throughout value chains. However, significant data gaps remain around the capacities and characteristics of women working in agriculture and agribusiness; the motivations of women entering into business; systematic analyses of entire value chains; and comparative studies.[22] Further research in these areas and tools such as the WEAI4VC can help clarify the gender dynamics of key sectors and products in a systematic way (Box 1).[23]

PRIVATE SECTOR INITIATIVE. Private sector institutions must also play an important role in making food systems more inclusive, given that food production, processing, transportation, trade, and consumption are driven by small, medium, and large enterprises. For instance, trade associations and certification initiatives can incorporate standards related to gender equality and women's empowerment. ISEAL–the global membership organization for credible sustainability standards–facilitates a Gender Working Group for Sustainability Standards to provide evidence-based strategies, tools, and systems to help standards organizations and multi-stakeholder initiatives integrate gender considerations and tackle systemic gender inequalities.[24] The group is focusing primarily on the textile and apparel sector, but a similar approach would be valuable in the agri-food sector.

CREATE AN ENABLING ENVIRONMENT. Incentives and regulations are needed to ensure that private sector investments benefit and empower women rather than exacerbate existing gender gaps in access to information, services, and other resources. Policymakers have an important responsibility to create enabling environments for research and industry developments that contribute to inclusive food system transformation. Having more women in leadership roles in all sectors will help to ensure that women's perspectives are included at the highest levels of influence.

DON'T LOSE GROUND. While there are clear opportunities to make food systems more equitable for women, it will be equally important to ensure that women do not lose ground as food systems transform. For example, as crops associated with women, such as cassava, commercialize, careful attention must be given to ensuring that women have the resources and decision-making power to expand production and take advantage of market opportunities.[25] Moreover, as food systems transform, it is important to monitor unintended consequences, such as increased workloads for women. For example, evidence from Mexico finds that although women who joined coffee grower organizations enjoyed increased decision-making within the home, they had a heightened perception of "time poverty" as a result of their involvement in the coffee schemes.[26] Evidence also suggests that approaches to empowering women must include working with men, both to prevent backlash against women's gains (such as gender-based violence or other retaliation) and to make sure that newly transformed gender norms are sustained.[27] Finally, just as institutions and technologies have the potential to empower women when planned deliberately, they can also exacerbate existing gender gaps. For example, deliberate steps must be taken to expand women's access to mobile phones and digital literacy to ensure they benefit from the potential of these innovations.[28]

Making food systems inclusive and gender-equitable requires recognizing women's roles and enhancing and ensuring opportunities for women to make strategic choices about their livelihoods, assets, relationships, and more. Transforming food systems to support and facilitate women's empowerment will benefit not only women but also their families and society.

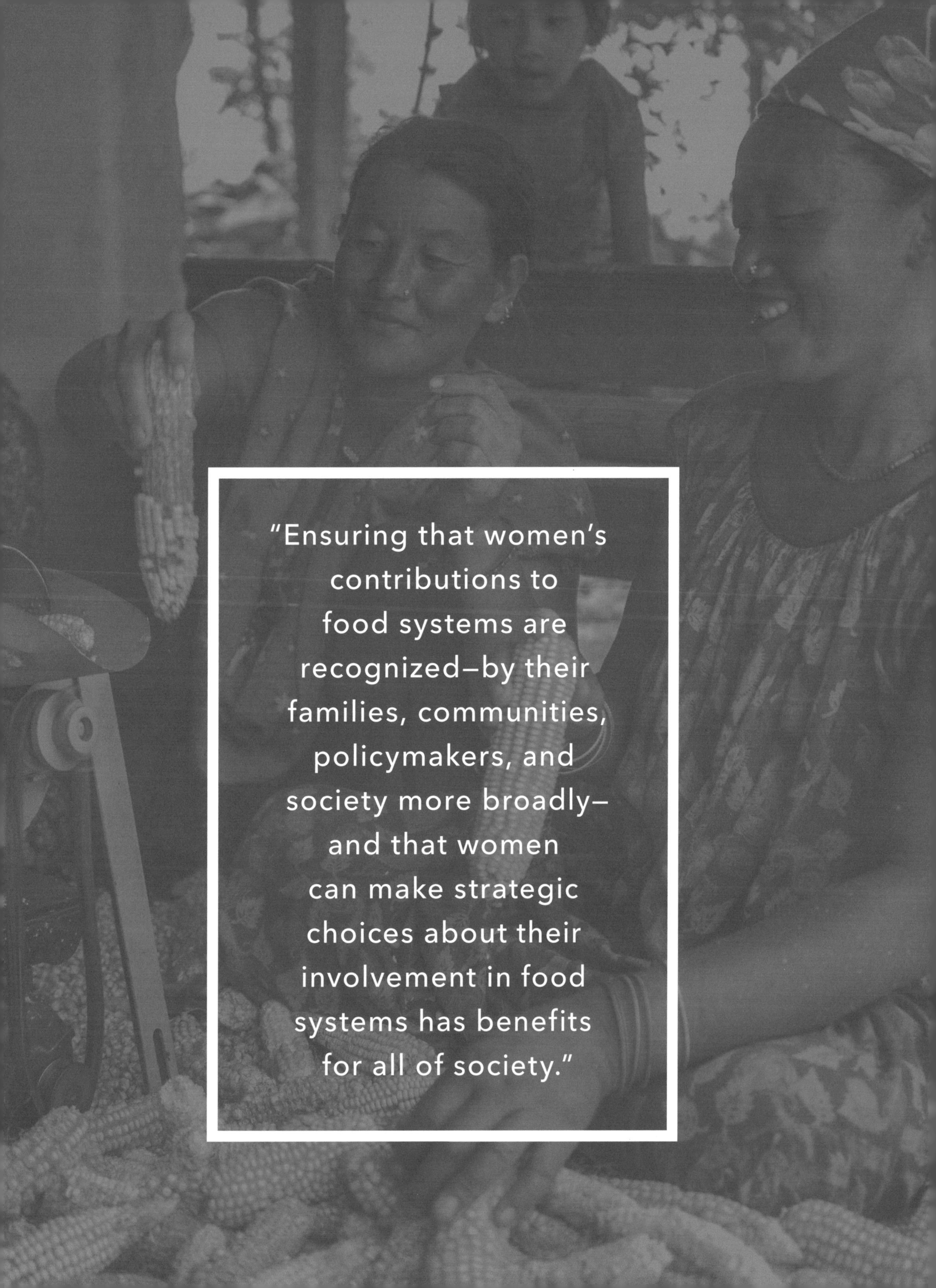

"Ensuring that women's contributions to food systems are recognized—by their families, communities, policymakers, and society more broadly—and that women can make strategic choices about their involvement in food systems has benefits for all of society."

CHAPTER 5

Refugees and Conflict-Affected People

Integrating Displaced Communities into Food Systems

ROB VOS, JULIUS JACKSON, SALLY JAMES, AND MARCO V. SÁNCHEZ

Rob Vos is the director of the Markets, Trade, and Institutions Division, International Food Policy Research Institute, Washington, DC, USA. **Julius Jackson** is a technical officer (protracted crises), **Sally James** is a forced migration and protracted crisis specialist, and **Marco V. Sánchez** is the deputy director, all in the Agricultural Development Economics Division, Food and Agriculture Organization of the United Nations, Rome, Italy.

KEY FINDINGS

- More than half of all undernourished people live in countries affected by conflict.
- Food insecurity and dispossession of agricultural assets can both trigger and result from civil strife.
- Most conflict-affected countries are overwhelmingly rural, and rural populations are more vulnerable to climate shocks that often compound conflict situations.
- Refugee host countries must often decide whether to focus responses on preparing affected populations to return home or helping them become economically self-reliant.
- Integrating conflict-affected people into food systems—either in their new homes or the places they fled—can help them rebuild their lives.

RECOMMENDATIONS

- Provide long-term refugees access to land and livelihoods to help them achieve food security while also strengthening local economies.
- Rebuild local agriculture and food value chains to help conflict-affected people move beyond subsistence agriculture, rejoin exchange markets, adopt climate-smart practices, and become resilient to economic and climatic shocks.
- Protect agriculture, food production, and rural livelihoods before, during, and after conflict.

Conflict and climate change have been key factors underlying the recent surge in global hunger numbers. People living in fragile rural contexts are most at risk. Food insecurity inflicted by conflict, climatic, and economic shocks, often in combination, is a main driver of forced migration and refugee flows. Such movements of people, whether within the borders of their own countries or crossing borders, often have adverse effects on food availability and access in host communities and areas left behind. Integrating conflict-affected people into food systems—either in their places of origin or the locales to which they have fled—could help them substantially to rebuild their lives. Furthermore, strengthening food systems and food security would remove at least one potential source of competition and conflict.

This chapter examines approaches and innovations to more fully include forcibly displaced and conflict-affected people (including host communities) in food systems, and the benefits of inclusion for these populations and society more broadly.

CRITICAL FACTS AND CHALLENGES

CONFLICT IS A MAJOR DRIVER OF FOOD INSECURITY. The recent rise in global hunger has resulted mainly from protracted conflicts: more than half of all undernourished people live in conflict-affected countries (Box 1).[1] Conflict is not the sole factor driving food crises. Drought, other climate shocks, and economic disruptions often are compounding factors. These same factors have also contributed to recent increases in forced migration. Every minute, 25 people flee their homes. UNHCR, the UN Refugee Agency, estimates that 70.8 million people were forcibly displaced in 2018, the highest number in decades (Box 1). Of these, 41.3 million were internally displaced, that is, they were forced to move to other localities within their own country, while 29.4 million were international refugees and asylum seekers (see Box 2 for definitions).

BOX 1 EIGHT INCONVENIENT FACTS

1. **CONFLICT IS THE NUMBER ONE DRIVER OF FOOD INSECURITY.** In 2016, 489 million of the world's 815 million undernourished people lived in conflict-affected countries.[a] Conflict, often compounded by climate change impacts, is also the main cause of food crises.[b]

2. **CONFLICT IS ALSO A MAJOR CAUSE OF THE GROWING REFUGEE CRISIS.** At the end of 2018, an estimated 41.3 million people were identified as being internally displaced because of armed conflict, generalized violence, or human rights violations, and 29.4 million as refugees or asylum seekers (see Box 2).[c]

3. **MOST REFUGEES MAINTAIN THIS STATUS FOR PROLONGED PERIODS.** In 2018, 15.9 million people had been refugees for five years or longer. Of this number, 5.8 million had been refugees in their host country for more than 20 years. Almost 3 million refugees currently live in camps.[c]

4. **AN ESTIMATED 600 MILLION YOUNG PEOPLE LIVE IN FRAGILE OR CONFLICT-AFFECTED AREAS.** 10- to 24-year-olds often comprise the largest group of the total affected population, yet little attention is paid to their needs or capabilities.[d]

5. **CHILDREN UNDER 18 REPRESENTED ABOUT HALF OF THE TOTAL REFUGEE POPULATION IN 2018.** The countries with the highest shares of young refugees are the Democratic Republic of the Congo (63 percent), South Sudan (62 percent), and Uganda (62 percent).[e]

6. **THE SHARE OF WOMEN AND GIRLS IN THE WORLD'S REFUGEE POPULATION WAS 48 PERCENT IN 2018.** However, the proportion varies from context to context. For instance, women make up well over half the refugee populations located in Africa south of the Sahara, while their share is smaller among refugees who have fled to Europe and South America.[e]

7. **IN MOST CONFLICT-AFFECTED COUNTRIES, THE MAJORITY OF THE POPULATION IS RURAL AND LARGELY DEPENDENT ON AGRICULTURE.** Agriculture's share in GDP averaged 37 percent in affected areas, two to four times higher than in developing contexts not affected by conflict or fragility.[f]

8. **UNDERNOURISHMENT IS HIGH IN CONFLICT ZONES.** In developing countries affected by conflict and crisis, the prevalence of undernourishment is almost three times higher than in other developing countries.[a,d,f]

FIGURE B1 Global forced displacement, 2009–2018

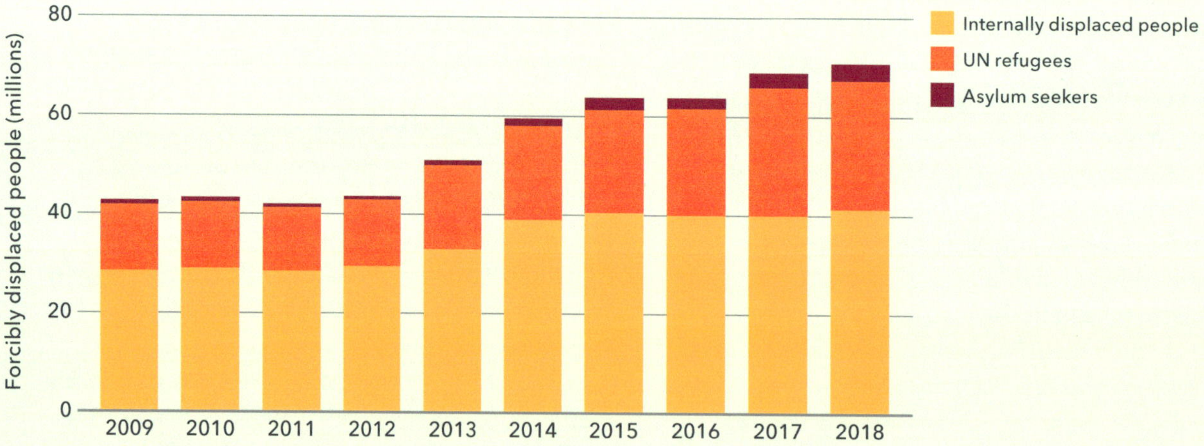

Source: UNHCR, *Global Trends: Forced Displacement in 2018* (Geneva: 2019).

Note: "UN refugees" includes UNRWA refugees (Palestine refugees registered with the UN Relief and Works Agency for Palestine Refugees in the Near East) and UNHCR refugees (all other refugees as counted by UNHCR, the UN Refugee Agency).

Source: (a) C. Holleman, J. Jackson, M. Sanchez, and R. Vos, *Sowing the Seeds of Peace for Food Security: Disentangling the Nexus between Conflict, Food Security and Peace*, FAO Agricultural Development Economics Technical Study No. 2 (Rome: FAO, 2017); (b) FSIN (Food Security Information Network), *Global Report on Food Crises 2019* (Rome: 2019); FSIN, *Global Report on Food Crises 2019 Mid-Year Update* (Rome: 2019); (c) UNHCR, *Global Trends: Forced Displacement in 2018* (Geneva: 2019); (d) FAO, IFAD, UNICEF, WFP, and WHO, *The State of Food Security and Nutrition in the World 2017* (Rome: 2017); (e) UNHCR, "Refugee Statistics," accessed December 5, 2019, https://www.unrefugees.org/refugee-facts/statistics/; (f) FAO, *The State of Food and Agriculture 2018: Migration, Agriculture and Rural Development* (Rome: 2018).

BOX 2 WHAT IS FORCED MIGRATION? WHAT ARE INTERNALLY DISPLACED PERSONS? WHAT IS A REFUGEE?

Forced Migration: A migratory movement in which there is an element of coercion, including threats to life and livelihood from natural or human-made causes. Forced migration can include movements of refugees and internally displaced persons (IDPs), but can also include people displaced by natural or environmental disasters, chemical or nuclear disasters, or famine, or as a result of large-scale infrastructure projects such as the construction of dams, roads, ports, or airports.[a]

Internally Displaced Persons (IDPs): People who have been forced to flee or leave their homes or places of residence as a result of, or in order to avoid, the effects of armed-conflict situations of generalized violence, violations of human rights, or natural or human-made disasters, and who have not crossed an internationally recognized state border.[b]

Refugee: Any person who, owing to a well-founded fear of being persecuted for reasons of race, religion, nationality, membership of a particular social group, or political opinion, is outside his/her country of nationality or former country of residence and is unable or, owing to such fear, unwilling to return to that country or avail himself/herself of the protection of that country.[c]

Source: (a) FAO, *FAO Migration Framework: Migration as a Choice and an Opportunity for Rural Development* (Rome: 2019); (b) United Nations, *Guiding Principles on Internal Displacement* (New York: 1998); (c) UNHCR, "Article 1A Refugee Convention 1951."

MOST CONFLICT-RELATED FOOD CRISES LAST MORE THAN 3 YEARS.[2] Likewise, internally displaced people (IDPs) and refugees typically remain in their new locations for prolonged periods. This creates a dilemma for host countries: should they focus responses on preparing affected populations to return home or on helping them become economically self-reliant and integrating them into their new communities? In most cases, security concerns limit options for safe and voluntary return and resettlement, while local integration may face significant resource constraints and opposition from host communities. These challenges are greatest for developing countries affected by conflict or their neighbors, as this is where 84 percent of refugees and IDPs are found.[3]

BOTH CONFLICT AND MASS DISPLACEMENTS OF PEOPLE DISRUPT FOOD SYSTEMS AND RURAL LIVELIHOODS IN COMMUNITIES OF ORIGIN, TRANSIT, AND DESTINATION. The impacts of conflict are felt across the entire food value chain, from production to marketing. Large influxes of people can also strain local food markets and basic services in communities hosting migrants.

CONFLICT, FORCED MIGRATION, AND FOOD CRISES

THE NUMBER OF CIVIL CONFLICTS AROUND THE WORLD HAS MORE THAN DOUBLED SINCE 2010.[4] Food insecurity can exacerbate feelings of deprivation and dispossession that underlie the recent rise in civil strife. In 2017, conflict in South Sudan caused famine in several parts of the country. In Yemen, home to today's worst humanitarian crisis, some 3.7 million people have been forcibly displaced since 2015, leaving more than 20 million food insecure and nearly 10 million on the brink of famine and starvation.[5]

AGRICULTURE-DEPENDENT PEOPLE AND RURAL DWELLERS ARE AFFECTED THE MOST, AS MOST OF TODAY'S CIVIL CONFLICTS TAKE PLACE IN SUCH CONTEXTS. Rural populations make up 60 percent of the total population of countries affected by conflict and protracted crises.[6] Rural populations are also more vulnerable to drought and other climate shocks that often compound conflict situations, destroying livelihoods and causing food insecurity. Syria's civil war, for example, has crushed the once-vibrant middle-income economy, leaving roughly 85 percent of the population impoverished, with more than 6 million people suffering from chronic hunger.[7]

In pastoral areas of Africa, protracted crises are harming livelihoods and disrupting longstanding livestock migration and trade routes.[8] Conflicts in Ethiopia, Kenya, and Uganda have contributed to the breakdown of traditional systems governing the mobility of herds seeking pasture and water.

PROTRACTED CONFLICTS ARE A MAJOR CAUSE OF THE RISE IN FORCED MIGRATION.[9] People living amid violence often must choose between the possibility of experiencing harm if they stay where they are or a highly uncertain and insecure future if they leave. Both options can be risky to survival itself. In Syria, conflict and drought have contributed to forced migration and displacement from rural areas, leaving fewer workers available for livestock-rearing and crop production. Many Syrian households have sold their livestock to generate income, as often to fund migration as to buy food, leaving them poorer and less resilient and weakening the local food system.[10]

FOOD INSECURITY AND DISPOSSESSION OF AGRICULTURAL ASSETS CAN ALSO TRIGGER CONFLICT.[11] While sudden food price spikes or loss of assets or harvests do not single-handedly cause conflict, they can stoke civil strife by compounding other grievances and discontent. In Yemen, for instance, overall economic decline, worsening living standards, and the government's failure to address these economic and social challenges have helped escalate political unrest into violent conflict.[12]

Food insecurity and outmigration also tend to disrupt social cohesion in local communities. Where governance is weak, rising food prices and food insecurity may result in perceived marginalization and exclusion, aggravating existing grievances.[13] Grievances formed along ethnic or religious lines (or other forms of social cleavage) increase the probability of civil unrest.[14]

THE REALITIES UNDERLYING CONFLICT, DISPLACEMENT, AND PERSISTENT FOOD INSECURITY TEND TO BE COMPLEX. Developing effective responses therefore requires a robust understanding of the root causes, and programs and support measures must address those root causes. Protecting agriculture, food production, and rural livelihoods before, during, and after conflict is crucial in most contexts. As food insecurity and conflict often feed one another, lasting solutions will be difficult to achieve if humanitarian, development, and peacebuilding responses do not come together.

ENGAGING DISPLACED AND CONFLICT-AFFECTED PEOPLE IN FOOD SYSTEM ACTIVITIES

Existing responses still consist mostly of patchy humanitarian and emergency actions supporting agricultural production, expanding social protection programs, and aiding displaced people in refugee camps and other areas of settlement. Yet policy assessments suggest the need for multiple well-coordinated responses that look beyond immediate emergency situations.[15] Accordingly, key actors have started reformulating their response frameworks along the "humanitarian-development-peace nexus."[16] That said, there is neither a prescription for how to address these problems nor any guaranteed remedy. But there are examples of promising responses that focus on strengthening food systems while helping to sustain peace by improving food security and resilience and allowing forcibly displaced people to take part in social and economic activities.

CLIMATE-SMART AGRIFOOD SUPPLY CHAINS CAN ENHANCE RESILIENCE TO CONFLICT AND OTHER SHOCKS. Any solution should take into consideration the needs of those who stay in affected regions, those who flee, and those belonging to host communities. Rebuilding local agricultural and food economies can help affected people to move beyond subsistence agriculture, rejoin exchange markets, adopt resilience-enhancing measures such as climate change adaptation, and stay in their community when it is safe for them to do so.

The post-conflict recovery in Uganda's Northern Region is a good example of how sustained investments in peace, recovery, and household resilience can lead to substantial improvements in food security and nutrition. Two decades of conflict between the Lord's Resistance Army (LRA) and government forces resulted in mass displacement and a surge in poverty and malnutrition, particularly among the Acholi people. Forced to live in camps, the Acholi became almost entirely dependent on international food assistance.

After the LRA's retreat in 2006–2007, IDP camps closed, with most residents returning to their places of origin in the following years. The Peace, Recovery, and Development Plan for the region facilitated peacebuilding efforts and prioritized investments in agriculture to cement post-conflict recovery. Multiple organizations have helped ex-combatants and returning IDPs get back on their feet through the provision of agricultural tools and inputs, including climate-resilient seeds, support for livestock restocking, and the introduction of cash- and food-for-work programs. These measures brought major improvements in food security and nutrition in the region; the Acholi have not needed any food assistance since 2011.[17]

PROVIDING LONG-TERM REFUGEES ACCESS TO LAND AND LIVELIHOODS CAN BENEFIT BOTH THE REFUGEES AND THEIR HOST COMMUNITIES. However, post-conflict political conditions often limit or preclude the possibility of third-country refugee resettlement or voluntary return and repatriation. For example, the many Rohingya refugees from Myanmar currently face bleak prospects in their new location in Bangladesh, given their severely limited income-earning opportunities (Box 3).

Yet, in some contexts, integrating large refugee populations into local economies generates brighter outcomes. Uganda currently hosts 1.2 million refugees, the third largest refugee population in the world.[18] The country's Refugee Policy (2006) and Refugee Regulations (2010) grant refugees access to land, freedom of movement, and the right to seek employment. This strategy has helped refugees, mostly from South Sudan, to build independent livelihoods and achieve food security while strengthening local economies.[19]

BOX 3 GRIM PROSPECTS FOR THE ROHINGYA IN BANGLADESH

Paul Dorosh (IFPRI) and John Hoddinott (Cornell University)

Between August and October 2017, 671,000 Rohingya fled violence and persecution in Myanmar for the safety of Cox's Bazar District in the Chittagong Region of southeastern Bangladesh. There, they joined Rohingya who had fled earlier violence over the previous 20 years. Fifty thousand of the early arrivals have refugee status, but the vast majority of the other approximately one million Rohingya are designated "Forcibly Displaced Myanmar Nationals (FDMN)," with no immediate hope of a safe return to Myanmar or a life outside the camps.

An October 2018 survey conducted by the Bangladesh Institute of Development Studies (BIDS) and IFPRI found that access to food assistance was nearly universal: 62 percent of the forcibly displaced Rohingya received a general food distribution consisting of rice, lentils, and micronutrient-fortified cooking oil; 34 percent received electronic food vouchers that could be redeemed for 19 different food items; and 4 percent reported receiving both.

Yet despite this international effort, the Rohingya are, at best, merely surviving. By any measure–income, consumption, assets–they are poor. While average caloric availability exceeds minimum required levels, Rohingya households consume very little fruit, vegetables, or animal-sourced foods, and 32 percent of children are chronically undernourished.

The Rohingya's long-term prospects may well be grim. To date, a return to Myanmar is neither safe nor viable, a view shared by both the Rohingya and UN humanitarian agencies. Meanwhile, other competing needs and donor fatigue threaten to reduce the level of support from both the Government of Bangladesh and the donor community. There are no easy solutions. In the short term, continued aid will be essential to avoid a humanitarian crisis. More lasting solutions will require the political resolve to provide the Rohingya the opportunity to rebuild their own sustainable livelihoods.

Source: This box draws on BIDS and IFPRI, *Forcibly Displaced Myanmar Nationals (FDMN) Household Survey, 2018* (Dhaka and Washington, DC: forthcoming).

Zambia and Ethiopia have applied similar policy approaches. The Zambian government has an official strategy for integrating Angolan and Rwandan refugees that regularizes their status and relaxes restrictions on freedom of movement. Likewise, the Ethiopian government recently introduced revised refugee-related laws and policies that extend a wide range of rights to Somali refugees, giving them access to services and land, helping them to establish new agricultural livelihoods, and facilitating social and economic integration into local communities.[20]

POST-CONFLICT SUPPORT TO RURAL RETURNEES IS CRUCIAL. Reviving the agriculture sector and improving livelihoods in post-conflict settings requires bridging humanitarian, development, and peace assistance. Stimulating the local economy, particularly in situations of protracted displacement, will help integrate migrants into the economy and the broader social fabric.

Such efforts are underway in post-conflict Colombia, where a half-century of armed conflict has inflicted severe social and environmental wounds. More than 8 million people were registered as victims and 7.8 million are still recorded as being internally displaced.[21] The conflict occurred mostly in rural areas, causing great loss of land and productivity, especially among small-scale producers.[22] The 2016 peace agreement includes a comprehensive plan for rural reform and revitalizing rural economies, providing services and benefits to victims, including land access to dispossessed and displaced farmers; improved infrastructure, agricultural practices, and natural resource

BOX 4 MITIGATING FOOD INSECURITY IN YEMEN

Yemen is in the throes of today's largest food insecurity crisis.[a] In January 2019, 16 million people, more than half the country's population, were in need of food, nutrition, and livelihood assistance, even when taking into account the mitigating effects of existing food provision and social protection programs. At that point in time, about 64,000 people in 45 districts suffered actual famine.[b] Over the course of the year, the situation for the most vulnerable in Yemen's worst-affected districts improved somewhat, thanks to scaled-up multisector assistance and greater food availability from seasonal harvests.[c] However, the food insecurity situation remains alarming in areas with active fighting, areas where internally displaced people (IDPs) and host communities face limited access to essential services and livelihood activities, and in the hard-to-reach areas that are home to 6.5 million people.[a,d]

Yemen's Cash Transfers for Nutrition program, which began in its current form in late 2016, provides cash transfers to pregnant women and mothers with children under two years old, provided they attend nutrition-focused trainings and comply with child nutrition monitoring programs. A recent IFPRI study found that the intervention had a positive impact on a range of intermediate outcomes, suggesting that the program both improved knowledge and increased spending on food.[e] In addition, the share of households benefiting from other food distribution programs almost tripled between 2015 and 2017.

During this period, average consumption of staple foods rose by 152 calories per person, despite a 23 percent rise in prices of imported dry staples. Yet dietary diversity decreased, as consumption of locally produced foods like vegetables and dairy declined. The average number of times each day that infants and young children were fed also declined.

While the Cash Transfers for Nutrition program was effective in raising calorie consumption and did contribute to a drop in the share of children diagnosed with moderate or severe malnutrition in the poorest households, it has been far from sufficient. Given the overall worrisome trends of worsening food insecurity and malnutrition in Yemen, more pervasive responses along the humanitarian-development-peace nexus will be needed, starting with reaching a peace settlement soon.

Source: (a) FSIN (Food Security Information Network), *Global Report on Food Crises 2019 Mid-Year Update* (Rome: 2019); (b) FSIN, *Global Report on Food Crises 2019* (Rome: 2019); (c) IPC (Integrated Food Security Phase Classification), "Acute Food Insecurity (Hot-Spot) Analysis," July-December 2019; (d) USAID, "Yemen—Complex Emergency," Factsheet no. 7, May 3, 2019 (Washington, DC); (e) S. Kurdi, Y. Ghorpade, and H. Ibrahim, "The Cash for Nutrition Intervention in Yemen: Impact Evaluation Study," IFPRI Middle East and Northern Africa Regional Program Working Paper (IFPRI, Cairo, 2019).

management; agrifood value chain development; employment programs; and social protection mechanisms. These programs aim to help conflict-affected families and returning IDPs to increase food production on family farms and restore local market infrastructure and social cohesion.

RISK-INFORMED AND SHOCK-RESPONSIVE SOCIAL PROTECTION SYSTEMS CAN HELP MITIGATE IMPACTS FROM FOOD CRISES AND ALLEVIATE PRESSURES TO MIGRATE. Such instruments aim to enhance household risk-management capacities and early responses to shocks and crises. Social protection systems are critical not only for short-term post-crisis relief, but also for preventing asset depletion at the household level and supporting asset creation at the community level.[23]

Cash transfer programs linked to agricultural production or nutrition-assistance programs (Cash+ programs) have proven effective in many conflict situations, helping affected households maintain access to food, avoid sale of assets, and strengthen household resilience. Although these programs can help avert the worst-case scenario, it is important to recognize that much more is needed to end the vicious cycle of conflict, displacement, food insecurity, and erosion of livelihoods.

Evidence shows that Yemen's cash-for-nutrition program, for example, has counteracted deterioration in food security and nutrition status as civil strife in the country intensified (Box 4).[24] Cash+ transfer programs in conflict-affected Mali and Mauritania, meanwhile, have been shown to improve incomes of beneficiary households and reduce the use of negative coping strategies, including selling land, deploying child labor, and begging, in response to adverse shocks.[25]

None of these examples should be taken as a blueprint for guaranteeing lasting peace, food system recovery, prevention of food crises, or restoration of livelihoods. But they do show that pathways toward such outcomes can be feasible if they build on a clear understanding of the root causes of conflicts, forced migration, and food crises and how they interact with and affect one another.

Most current conflicts are fought in rural areas and cause severe food insecurity. Resulting stresses are likely to fuel further conflict and force people to flee. Humanitarian interventions that have the greatest likelihood of achieving lasting success involve investing in local agrifood systems and including conflict-affected people in strategies and programs for building, reviving, or strengthening these systems.

CHAPTER 6

National Food Systems
Inclusive Transformation for Healthier Diets

JOHN MCDERMOTT AND ALAN DE BRAUW

John McDermott is the director of the CGIAR Research Program on Agriculture for Nutrition and Health (A4NH), International Food Policy Research Institute (IFPRI), Washington, DC, USA. **Alan de Brauw** is a senior research fellow in the Markets, Trade, and Institutions Division, IFPRI.

KEY FINDINGS

- The rapid transformation of national food systems offers new opportunities for inclusion of poor and marginalized people, potentially improving dietary diversity, food safety, and quality.
- As food systems transform across the spectrum from traditional to modern, government policy goals need to shift from a focus on food security to healthy, balanced diets.
- National food system frameworks are useful tools for looking at the drivers and components of these systems, identifying data gaps, and finding promising entry points for actions to increase inclusion and improve nutrition outcomes.
- Approaches to food system transformation must be country specific, as each country's food system is unique and countries face different opportunities and trade-offs for inclusiveness at different stages of transformation.

RECOMMENDATIONS

- Reverse traditional thinking about food systems by starting from the consumer, focusing on diets and consumer demand. Better collection of data on changing diets, especially consumption of processed foods, and development of nationally appropriate dietary guidelines can inform strategies to address rising obesity and persistent malnutrition.
- Combine technological innovations, institutional capacity, and infrastructure investments—such as use of information and communications technology, food quality certification, and cold chains—to catalyze positive systemic change at the national level.
- Continually adapt policies as food systems evolve to ensure they promote healthy diets, create an enabling environment for positive private sector contributions to making food systems inclusive, and manage trade-offs among different policy goals.

National food systems in low- and middle-income countries (LMICs) are transforming rapidly from traditional to modern. This is part of a larger story of rural transformation, urbanization, and development that can offer new opportunities for inclusion of poor and marginalized people. In addition to creating employment and income-generating opportunities, transformation can also support improvements in nutrition that are associated with long-term impacts on health, cognitive capacity, educational attainment, income, and development. The tools and policies for making food system value chains more inclusive have been described in the previous chapters, as have the particular obstacles and opportunities facing smallholders, women, youth, and refugees. Here we identify some of the challenges of ensuring that national food system transformations contribute to better diets and nutrition outcomes for all.

As countries urbanize and incomes rise, consumers begin to demand more diverse, convenient, and safe foods. Changes in food demand can drive changes throughout a food system, ranging from farmers to supply chains, markets, and households. Yet poverty, geographic isolation, gender, and other inequalities can exclude people from new opportunities created by national food system transformations and can make healthy food difficult to access. These failures are evident globally in a range of indicators: disappointingly slow reductions in rates of child stunting, persistent hunger, stubbornly high prevalence of micronutrient deficiencies, and rapidly increasing rates of overweight and obesity. Many countries are dealing with several of these, and often all at once. Policymakers need to know what policies, investments, and actions they can take to ensure food systems transform in a healthy, sustainable, and equitable way.

Informed policymaking will require a better understanding of how food systems affect nutrition, what entry points and policies are most effective, and what trade-offs must be made. This chapter reviews the framework and findings of the Food Systems for Healthier Diets research program under the CGIAR

The authors thank Inge Brouwer, associate professor, Division of Human Nutrition and Health at Wageningen University & Research (WUR) and A4NH Food Systems for Healthier Diets flagship leader, and Ruerd Ruben, professor and research coordinator at WUR and an A4NH managing partner representative, for their insights on food systems from a national perspective. The authors also thank Victor Manyong, agricultural economist at the International Institute of Tropical Agriculture (IITA), for his contributions, and Janet Hodur, A4NH senior communications specialist, for her support in writing this chapter.

FIGURE 1 Stages of food system transformation

TRADITIONAL

- Policy focus on food security and supply and cereal production
- High share of agriculture in GDP (>25%) and employment
- Food is eaten close to where it is grown
- Sustainability focus on climate adaptation and efficiency and effectiveness of land and water use
- Limited industry (packaging, processing)
- Most calories from cereals (>75%)
- High levels of stunting (>35%)

TRANSITIONING

- Policy focus expands to consider micronutrient intake, dietary diversity, and agricultural transformation
- Decreasing share of agriculture in GDP (10-25%) and employment
- Increasing rates of urbanization
- Increasing lengths of food supply chains and increasing food safety burden
- Sustainability issues more complex, with greater options for specialization and market-based solutions
- Industry (packaging, processing) begins to grow
- Decreasing share of calories from cereals (65-75%)
- Variable levels of stunting (25-50%)

MODERNIZING

- Policy focus on food system transitions, food quality and safety (although food safety burden falls)
- Low share of agriculture in GDP (<10%) and employment
- High rates of urbanization
- Complex sustainability synergies and trade-offs, benefit from systemic approach to food systems decision-making
- Industry plays large role
- More trade, with greater share of food imported and exported
- Food supply chains are more integrated and complex
- Lower share of calories from cereals (<65%)
- More food is eaten away from home (snacks, restaurants, etc.)
- Moderate levels of stunting, still declining (<30%)

Research Program on Agriculture for Nutrition and Health (A4NH).[1] This program is engaging with national-level development practitioners, entrepreneurs, and policymakers to develop evidence on national food system transformation in four focus countries—Nigeria, Ethiopia, Bangladesh, and Viet Nam—and on subnational food systems in India, in order to assess possible system interventions and enabling actions to scale and anchor desired food system outcomes.

A NATIONAL PERSPECTIVE ON FOOD SYSTEM TRANSFORMATION

Food system transformation is now central to the development strategies of most LMICs. This emphasis reflects the need to meet growing domestic and global demand for food resulting from growing populations and rising incomes. It also reflects the recognized potential for food system transformation, both to provide more and better employment and value addition beyond primary agriculture *and* to improve nutrition through better access to healthy diets. But few countries have developed specific plans for food system transformation or engaged the coalition of public and private organizations necessary to implement them.

Country-specific approaches are needed because each country's food system is unique, reflecting national natural resources, market access, and sociocultural traditions as well as the country's stage of economic transformation. We categorize food system transformation into four stages—agrarian or traditional, transitioning, modernizing, and modern.[2] The first three stages predominate in LMICs. Figure 1 describes the characteristics common to each of these stages.

As systems evolve from one stage to the next within any given country, policies need to change and adapt. Designing appropriate investments, policies, and regulations to include, enable, and incentivize food system actors is particularly challenging when the food system itself and strategic priorities are rapidly evolving. In LMICs, government policies and goals commonly progress from an initial food security focus (having enough to eat) in traditional systems, to basic diet adequacy (ensuring adequate micronutrient intake) as countries transition, and then to healthier balanced diets (access to safe, healthy, and diverse foods) as they modernize. Managing these transitions to ensure greater inclusion and better nutrition outcomes has potential to positively change the trajectory of health, equity, and sustainability.

The focus countries of our research program have important differences but share key policy objectives that will contribute to improvements in nutrition for all. These countries all aim to enable value addition beyond the farm for greater economic growth and jobs, which will raise incomes for many. They also aim to diversify food supply chains to increase the availability of nutrient-dense foods, such as fruits, vegetables, and animal-sourced foods, and to limit consumption of less healthy foods high in sugar, fats, and salt in order to improve nutrition outcomes. Depending on the country context, different approaches will be more effective in reaching these goals.

FOOD SYSTEM FRAMEWORKS

Food system frameworks are useful tools for assessing how food systems are contributing to reducing malnutrition in all its forms, and whether they are doing so in an equitable and sustainable way. We use a simplified version of a widely accepted national food system framework to help carry out a three-step analysis that will allow national actors to build a food system strategy to meet their food system and nutrition objectives.[3] These steps are:

1. Assess the country's development context, particularly demand drivers shaping food systems and how these will change in 10 and 20 years.
2. Understand what indicators are available to assess health, sustainability, and equity outcomes.
3. Identify priority areas for action for the food supply, food environment, and consumer behavior.

Figure 2 depicts our framework's key elements: a broad set of *drivers* that influence three food system *components*—namely, value chains, the food

FIGURE 2 Simplified framework for food system analysis

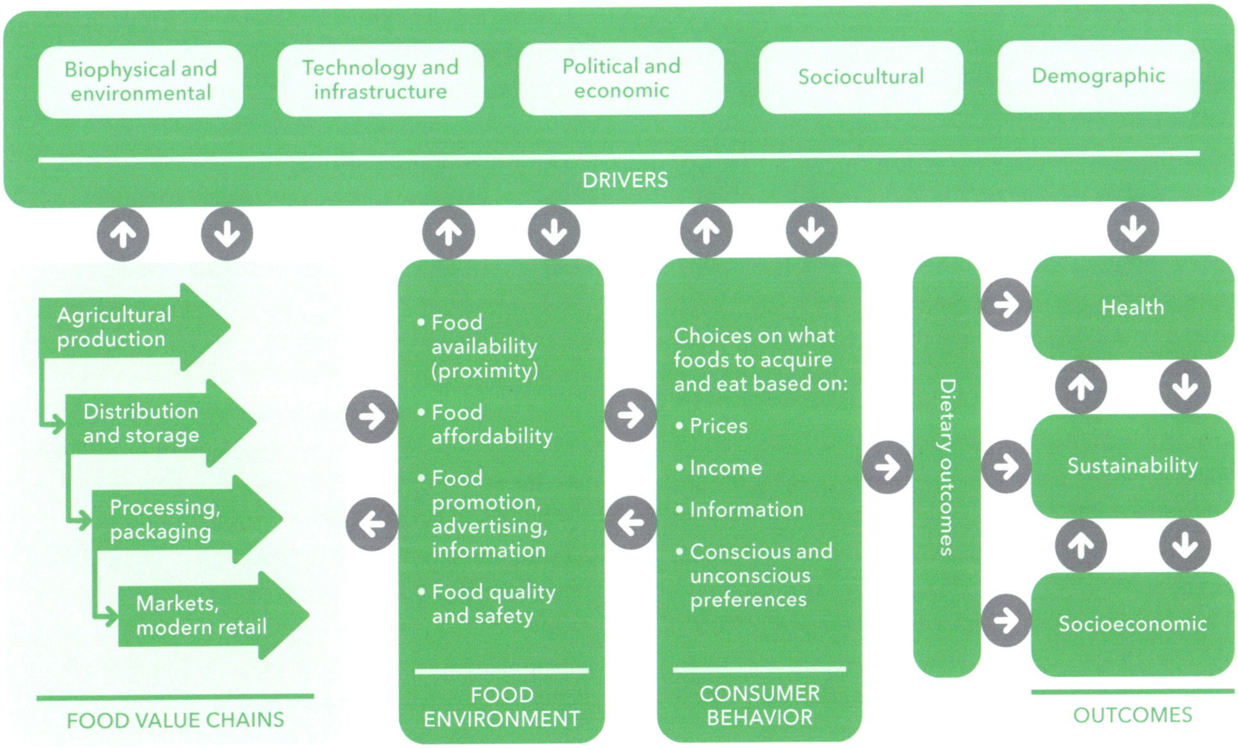

Source: Adapted, with permission, from HLPE (High-Level Panel of Experts), *Nutrition and Food Systems: A Report by the High-Level Panel of Experts on Food Security and Nutrition* (Rome: 2017).

environment, and consumer behavior. Together these determine socioeconomic, health, and sustainability *outcomes*. Central to this framework is the understanding that food systems are demand-led. The choices people make about diets are shaped by prices, income, information, and marketing and by individual needs, preferences, and constraints—and these choices determine nutrition and health outcomes. The dynamics of the food system also shape sustainability and equity outcomes. Looking at the drivers and food system components for a particular country through this framework can help policymakers identify policy needs and promising leverage points for action. It can also highlight systemic trade-offs, interactions, and synergies in achieving nutrition, sustainability, and socioeconomic outcomes.

For most countries at the traditional or transitioning stages of food system transformation, critical information gaps exist regarding food system impacts and outcomes. This framework provides a useful way to visualize the availability of relevant indicators and data gaps (Box 1).

INCLUSION STRATEGIES FOR FOOD SYSTEM TRANSFORMATION FOR HEALTHIER DIETS

Four national strategies for inclusive change in food systems offer promise for improving nutrition among disadvantaged groups, including poor populations and other disempowered social groups, such as women and refugees. In discussing each strategy, we look at country experiences, drawing heavily on research in the four focus countries and Indian states. At the different stages of transformation—traditional, transitioning, and modernizing—different approaches will be needed to implement these strategies. These evolving policy needs are summarized in Table 1.

REVERSE THINKING: *Put diets first*

Traditional and transitioning countries usually focus on increasing the food supply, but not on consumer behavior or food environments.[4] Reversing this thinking by starting from the consumer demand end of a food system can help address growing problems associated with unhealthy diets, including the rise in consumption of highly processed foods, that may disproportionately affect children and the poor.

One entry point is to develop food-based dietary guidelines suitable to the national context. National dietary guidelines must translate existing scientific knowledge on the links between foods, diets, and health outcomes into local food habits and dietary patterns, food availability, and costs. In Ethiopia, a technical working group, composed of representatives from government ministries, NGOs, academia, and civic organizations, is using a consultative process to develop national guidelines based on scientific evidence and local conditions and preferences. These guidelines can be used by consumers to inform food choices and by policymakers for formulating policies and strategies. Development of these guidelines has stimulated discussion of diet and nutrition issues. The process has benefited from strong government interest, and Ethiopia's experiences are being shared with other countries.[5] In parallel, Ethiopia is linking nutrition and equity through the Seqota Declaration program. This multisectoral program is combining agrifood, health, water and sanitation interventions, and social protection programs in districts where childhood stunting levels are high. Together these efforts are moving Ethiopia beyond an earlier and singular focus on food security toward a broader food system and nutrition approach.

As countries begin to shift away from traditional diets, foods eaten away from home (snack, street, and restaurant foods) are consumed with increasing frequency, often with deleterious nutrition and health impacts.[6] For example, in India snack foods are aggressively marketed and widely consumed by children and adolescents, and obesity problems and associated diabetes and cardiovascular diseases are rising at an alarming rate (Box 2).[7] Yet the impacts are not consistent across countries: the rise in overweight and obesity in South Asia and Africa south of the Sahara has been greater in wealthier households and urban areas, while poorer and more rural households have faced the greatest increases in other LMICs.[8]

Capturing information on food consumption outside the home and incorporating it into dietary guidelines and policy actions poses a challenge for traditional and transitioning-stage countries. Current surveys of food consumption expenditures are not designed to track these purchases, and the lack of data is a major concern for countries that need to

BOX 1 INDICATORS FOR ASSESSING FOOD SYSTEMS

Inge Brouwer (Wageningen University & Research) and Alan de Brauw (IFPRI)

As part of the UN Sustainable Development Goals (SDG) process, all countries must collect data for SDG indicators. For assessment of food systems, the SDG indicators need to be supplemented with other available data. Table B1 illustrates how the food system framework can be used to look at the current state of publicly available information for national and subnational food system assessment and decision-making across countries. In those African countries where food systems are traditional or transitioning, most policy emphasis is on food supply, and the availability of indicators reflects this emphasis. Nigeria and Ethiopia, for example, focus on agricultural transformation, reflected in more indicators for agricultural productivity. Most countries have data and indicators for their relatively small processing and packaging industries, but information on logistics, storage, and marketing is quite limited, as are data on food environments. Data availability does not always indicate stage of development, however: Bangladesh has substantial data on food environments while, surprisingly, Viet Nam does not–recent innovations and investments in Viet Nam are not yet reflected in publicly available data. Generally, data on national food systems reflect the persistent emphasis on production and do not necessarily represent dynamic change in the system. Reorienting thinking and data collection toward determinants of food demand, especially the food environment component, is critical to managing the diet transition.

TABLE B1 Information and data available to assess food system indicators in Bangladesh, Ethiopia, Nigeria, and Viet Nam

	INDICATORS AVAILABLE (Substantial / Some / Few or none)			
	Bangladesh	Ethiopia	Nigeria	Viet Nam
Food value chains				
Agricultural production	Some	Substantial	Substantial	Some
Distribution and storage	Some	Few or none	Few or none	Few or none
Processing, packaging	Substantial	Some	Substantial	Substantial
Markets, modern retail	Few or none	Few or none	Few or none	Few or none
Food environment	Some	Few or none	Few or none	Few or none
Consumer behavior	Some	Substantial	Substantial	Substantial
Drivers				
Biophysical and environmental	Substantial	Substantial	Substantial	Substantial
Technology and infrastructure	Few or none	Few or none	Few or none	Few or none
Political and economic	Substantial	Substantial	Some	Substantial
Demographic	Substantial	Substantial	Substantial	Substantial
Outcomes				
Dietary and health	Substantial	Substantial	Substantial	Substantial
Sustainability	Few or none	Few or none	Few or none	Few or none
Socioeconomic	Few or none	Few or none	Few or none	Few or none

TABLE 1 Examples of inclusive policies and actions, by type and transformation stage

TYPE OF REFORM	STAGE OF FOOD SYSTEM TRANSFORMATION		
	TRADITIONAL	TRANSITIONING	MODERNIZING
REVERSE THINKING: PUT DIETS FIRST	Promote production of nutrient-dense foods	Nutrition education or information campaigns around healthy eating	Require packaging labels (or QR codes) and labeling of foods eaten away from home
FOOD SYSTEM INNOVATION (TECHNOLOGY, INSTITUTIONS, AND INFRASTRUCTURE)	Biofortification; contractual innovations in nutrient-dense food production	Food quality certification; cold chain innovations; technological agricultural extension	Infrastructure for logistics to enhance efficiency and traceability
ENABLING POLICY ENVIRONMENT (PRIVATE SECTOR, CEREAL SUPPORT REFORMS, REGULATION)	Food fortification in processing; combine social protection with dissemination of information on diets	Develop food-based dietary guidelines; implement soda or unhealthy food taxes; purchase of nutrient-dense foods by schools and institutions	Healthy food subsidies; food safety regulations
ENSURING INCLUSIVITY OF INNOVATIONS AND REFORMS	Target smallholders with biofortification; ensure contract terms are incentive-compatible for all parties; include farmer groups and extension in contracts; understand impacts of any regulation and innovation for gender or vulnerable groups	Ensure that poor farmers trust quality certification, whether done by government or third party; target tax revenue for health among poor; technology penetration to poor necessary for extension to be effective	Provide vouchers for nutrient-dense foods among poor; consider effects of regulations on food costs among poor; require labels (or QR codes) on all packaged foods

BOX 2 SNACK FOODS AND IMPLICATIONS FOR CHILDREN AND ADOLESCENTS IN PUNE, INDIA

Anjali Ganpule (Gokhale Research Institute), Bhushana Karandikar (Gokhale Research Institute), Avinash Kishore (IFPRI), Devesh Roy (IFPRI), and Manika Sharma (IFPRI)

In India, traditional diets and eating habits are changing in response to rising incomes, and snack foods have become an important part of the food environment. To investigate eating outside the home, we conducted a cross-sectional study of 1,500 people along a rural–urban transect in Pune, Maharashtra State. Snack food consumption was common in the country's transitioning and modernizing food systems. Children ate snack foods 2–3 times per day; adolescents ate snack food 1–2 times per days. Among adults, urban residents snacked more.

Snacking trends in Pune mirror the increase in processed food consumption in other transitioning and modernizing food systems. People's snack food choices were largely driven by price and taste, with little attention paid to ingredient labels, and snacks high in sugar, salt, and low-quality fat were most popular. Children and urban consumers were more likely than rural adults to be influenced by brand advertising.

Our assessment of weekly food consumption showed that unhealthy and cheap snack foods are crowding out healthier foods, including pulses, coarse grains, and vegetables. Given the obesity epidemic (in this survey, more than 50 percent of urban adults were obese and in rural areas 29 percent of women were obese) and the high rate of associated noncommunicable diseases in India, efforts to encourage healthier eating, such as the Eat Right India campaign, are increasingly important to ensure a healthy food environment for children and adolescents.

Note: For more information, see A. Ganpule-Rao, A., D. Roy, B. Karandikar, C. Yajnik, and E. Rush, "Food Access and Nutritional Status of Rural Adolescents in India: Pune Maternal Nutrition Study," *American Journal of Preventive Medicine* (forthcoming, 2020).

manage the increasing consumption of unhealthy snack foods. Globally, little progress has been made in slowing the rise of obesity, due to lack of concern from consumers, resistance from food companies, and limited incentives for national political leadership on the issue.[9] While experience with policies, regulations, and interventions to fight obesity is growing, including in LMICs, there has not been a systematic way of cataloging this information.[10] Some countries at the modernizing stage of food system transformation, notably Chile and Mexico, have taken actions such as nutrition labeling and soda taxes, respectively. These innovations could add to the tool kit for traditional and transitioning countries to ensure food system transformation doesn't lead to unhealthy diets and worsening nutrition.[11]

FOOD SYSTEM INNOVATION:

Combine technology, institutional capacity, and infrastructure

As a country begins to shift out of a traditional food system, innovations in technology, institutions, and infrastructure can work together to contribute to positive, inclusive systemic change. For food supply transformation for healthier diets, combinations of technology, institutional change, and enabling policies have been critical in areas such as rice production in Bangladesh, labeling of foods in Viet Nam, and the transformation of India's dairy industry. In the Indian case, dairy cooperatives began by helping smallholders with feed, loans, and other inputs as well as links to markets for their perishable milk. Over time, the cooperatives developed capacity for production of cheese, yogurt, and other high-value products, which has opened up new opportunities.

Technological innovations that drive agricultural transformation can be adapted for broader impact on food system performance for nutrition. Information and communications technology (ICT) innovations are most successful when they fit farmers' needs and when farmers' trust in the system supports a critical mass of users, allowing for network effects. Most notably, ICTs show promise for increasing inclusion: ICTs can improve production and improve access to rural services, credit, and market information (see Chapter 2).[12] They can also be adapted to trace food from farm to consumer, helping assure consumers of food safety and quality, which contribute to healthy diets.

In our focus countries, two *institutional-strengthening approaches* show promise for reducing common inefficiencies in the production and supply of nutrient-dense, perishable foods such as fresh fruits and vegetables and animal-sourced foods. First, farmers' organizations can act as aggregators for knowledge and help link small-scale farmers with both input and output markets (see Chapter 2), as in the Indian dairy example. Second, quality certification can allow smallholders to charge higher prices for products that meet consumers' demand for quality, such as certified organic foods, or other standards, such as size and safety.[13] Where governments are considered reliable, they can provide certification; otherwise, third parties can provide this service. Even assurance that goods can be tested by third parties can improve prices paid to smallholders.[14]

Infrastructure investments can also contribute to inclusive growth in food systems and diversification of the food supply—both healthy and unhealthy. Roads are known to have broad general effects on inclusiveness by facilitating linkages: trunk, or major, roads have been shown to increase agricultural trade and income; however, evidence on trade and income impacts of feeder roads is less clear.[15] Cold chains are critical for expanding markets for smallholders' high-value perishable products and for delivering these nutritious goods to urban consumers. Investment in cold chains has largely been led by the private sector.[16] Several cold chain innovations are being developed, such as the CoolBot technology, which uses standard air-conditioning equipment to create a cold storage space and has low fixed costs, and the Dearman engine, which uses liquid air as fuel to cool spaces such as trucks but requires substantial upfront investment. While both of these technologies can potentially deliver more perishable goods to urban areas at lower costs, rigorous assessment of their nutrition and equity impacts is needed.

ENABLING POLICY ENVIRONMENT:

Let markets work for inclusion and healthier diets for all

Providing an enabling environment for effective and inclusive food systems is challenging, and policies often produce unintended consequences, both positive and negative. Developing countries must be able to adapt and change policies to reflect their changing circumstances and must design food system policies with a view to likely trade-offs to balance health, sustainability, and equity outcomes for overall welfare.

As countries transition toward modern food systems, they must manage the shift in policy objectives from food security to food quality. In traditional food systems, domestic cereal production for food security dominates national concerns. Since the 1970s, a policy emphasis on increasing production of staple foods has contributed to the global decline in cereal prices in real terms, relative to much higher prices for nutrient-dense foods, such as vegetables and fruits, pulses, and animal-sourced foods.[17] In many LMICs, the absolute price of nutrient-dense foods is significantly greater than in high-income countries.[18]

A range of policies and investments that drive the relative prices of staple foods have shaped diets in many countries. Markets for food can be distorted by public policies such as subsidies, as was the case with subsidized water and energy in India and fertilizer in Malawi; investment in research and development, as for maize in Malawi and Zambia; import tariffs, as for rice in Nigeria; and bans and export restrictions in many countries. These policy-induced distortions can have positive or negative impacts on diets, depending on the national context. In India, keeping the cost of cereals low relative to noncereals has contributed to the proliferation of cheap snack foods.[19] But in Bangladesh, the stable supply of low-cost rice, combined with economic growth, has allowed the poor to increase consumption of nutrient-rich fish and vegetables.[20] The challenge is to make policies inclusive and forward-looking—moving from a focus on food security to a focus on a diverse food supply and postfarm food processing, logistics, and market investments—as food systems transform and national needs change.

Small and medium-sized enterprises (SMEs) can play a vital role in making food systems more inclusive by creating jobs and linking rural and urban areas (see Chapter 2). Policies and regulations are particularly important in creating an enabling environment for inclusive private sector food system activities. In our focus countries, private sector actors at all stages of the food supply chain are largely SMEs, including for-profit and social enterprises, cooperatives, and farmer-producer organizations. SMEs often lack essential technical skills and business experience, and because of their small size are unable to bear much risk, making scaling up these operations a critical challenge.[21]

In Africa and Asia, there is considerable interest in fostering private sector involvement in food system transformation. Policymakers embrace the concept of food system transformation because of the potential for increasing inclusive growth, benefiting women (see Chapter 4) and youth (see Chapter 3) particularly. The agriculture and food sectors have the potential to provide jobs for youth, if supported with increased investment and conducive legal and policy environments.[22] Agribusiness SMEs—that is, farming plus all the industries and services along the food value chain—could generate opportunities for youth in food processing, wholesaling, and retailing.[23] Given the demographic change in Africa, where the "youth bulge" raises both opportunities and challenges, African countries are particularly keen on boosting youth employment through food system development.

Developing appropriate policies and regulations to balance the critical food system outcomes—that is, managing trade-offs—is challenging in all countries. As food systems become more modern, demand increases for higher food quality and for food safety, prompting governments to adopt stricter food standards. But often in LMICs, these policies and regulations are too complex to be implemented or they disadvantage those less able to participate in modern value chains.[24] Among our four focus countries, the transformation of food systems has been most rapid in Viet Nam, which has been at the forefront of the debate over restrictive food safety standards required by export markets and supermarkets and the less stringent food safety regulations appropriate for local fresh food markets, where most poor people buy and sell foods (Box 3). The country is trying to manage trade-offs to ensure that food safety governance and regulation do not exclude the poor as sellers and buyers of fresh foods, while also promoting growth in higher-value domestic and export food markets. At present, a compromise allows informal markets to function while commercial food systems must meet more stringent quality and safety standards. Food labeling for nutrition and safety is now mandatory for the commercial food system in Viet Nam, and healthier foods are marked by a special label developed by an NGO.[25] This is part of a small but rapidly growing international body of experience in policies and regulations for mitigating obesity.[26]

BOX 3 MANAGING FOOD SAFETY GOVERNANCE IN VIET NAM

Hung Nguyen and Delia Randolph (International Livestock Research Institute)

The switch from "needing enough food to eat" to "needing safe food" is happening rapidly in Viet Nam. With more than 80 percent of food purchased in traditional "wet" markets, food safety management is an important issue. The poor, women, and children are the groups often exposed to food hazards from the more than 9,600 traditional markets (compared to 700 supermarkets) and many other informal markets across the country that supply most fresh, perishable foods. Finding appropriate ways to manage food safety for these markets without decreasing employment opportunities and access to fresh food for the poor is critical.

At the same time, Viet Nam is already a major exporter of high-value food (including seafood, vegetables, and rice), and its exporters are meeting the demanding requirements of US and European markets. Export value chains benefit from substantial private sector capacity, knowledge, and protocols for producing safe food, and people working in exports have substantial training and earn a notable premium.

The government is committed to establishing a single standard for both exports and domestic food to replace the lower standards that prevail in wet markets. But the implementation of food safety management in the large system of informal markets is weak and marked by poor compliance in production, processing, and marketing. And when food scandals inevitably occur, public mistrust and miscommunication about food safety risks are amplified through social media. During this transition period–when value chains for supermarkets are long and complex and those for wet markets are local and simple–food systems are evolving toward developing globally recognized quality and safety standards.

Improving domestic food safety in Viet Nam will require capacity building, effective verification of safety, and incentives for value-chain actors, including rewards for safe food and penalties for unsafe food. Fortunately, the 2011 Food Safety Law provides a modern framework for Viet Nam's food safety management system. Its implementation will require increased capacity across government levels to apply a risk-based approach for both export and domestic food markets, and will allow lessons from the export sector to be shared for the benefit of the domestic sector.

Note: For more information, see H. Nguyen-Viet, T. T. Tuyet-Hanh, F. Unger, S. Dang-Xuan, and D. Grace, "Food Safety in Vietnam: Where We Are At and What We Can Learn from International Experiences," *Infectious Diseases of Poverty* 6, no. 1 (2017): 39; World Bank, "Food Safety Risk Management in Vietnam: Challenges and Opportunities," Technical working paper (World Bank, Hanoi, 2017).

SOCIAL DEVELOPMENT POLICIES:
Including people and places left behind

Policies tailored to national conditions must also take account of the people and places left behind by transitioning and modernizing food systems. Many national and subnational food system interventions have not been sufficiently inclusive nor contributed sufficiently to equity. In this section, we describe how countries have implemented supplemental social development policies and actions intended to complement and enhance inclusive food system transformation.

Research and policy advice are paying greater attention to those excluded or disadvantaged by changing food systems in terms of access to food, food quality, and the other benefits of more modern food systems. Poverty, nutrition, and health outcomes have been shown to be related to social groupings (such as tribe or caste), climate emergencies, natural resource degradation, and conflicts (see Chapter 5).[27] In addition, understanding of the impact of gender roles and gender empowerment on inclusion and nutrition is expanding (see Chapter 4).[28] Given a growing body of evidence on poor food system outcomes, interest is growing in more radical approaches to supporting groups being left behind.[29]

In Ethiopia, for example, interventions to increase agricultural productivity supplemented by social development actions to improve food security and nutrition for groups left behind have helped reduce food insecurity across both the country's productive agricultural zones and its poor drought-prone zones (Box 4). In India, deliberate efforts have been made to link nutrition and inclusive growth in lagging districts, supported and incentivized with a unique outcome-oriented governance approach (Box 5). And in Malawi, the positive nutritional impact of both food transfers during the lean season and behavior change communication to support dietary diversity suggest that combining the two approaches could boost healthy diets for the disadvantaged (Box 6).

BOX 4 ETHIOPIA SUPPLEMENTS GROWTH PROGRAM WITH SOCIAL SAFETY NET

Kalle Hirvonen (IFPRI) and John Hoddinott (Cornell University)

Agricultural output in Ethiopia has more than doubled over the past decade, but the spatial distribution of gains remains highly uneven. While the western highlands enjoy near-optimal climatic conditions for agricultural production, the eastern region is subject to frequent droughts, which leave its population chronically food insecure.

The government of Ethiopia, together with a consortium of international donors, provides targeted investments appropriate to these two different areas. Through the Agriculture Growth Program (AGP), the high-potential areas receive support to improve agricultural productivity and market performance of crop and livestock value chains, allowing the region's smallholders to take advantage of its natural resources and market connections. The AGP currently covers 157 districts and 1.3 million smallholder farmers. Complementing it, the Productive Safety Net Program (PSNP) provides transfers to address chronic food insecurity in drought-prone areas. The program is currently implemented in more than 300 districts, with more than 8 million beneficiaries, making it one of the largest safety net programs in Africa. Between 2006 and 2014, food security improved considerably among households that took part in the public works component of the PSNP, and about 80 percent of this improvement can be attributed to the program. Evidence showing limited improvements to child nutritional status led to a redesign of the program in 2014–2015 to incorporate nutrition-sensitive components that combine poverty reduction, food security, and nutritional benefits. In addition, efforts to link the PSNP to interventions aimed at increasing agricultural output have led to increased fertilizer use and agricultural investment in PSNP districts.

Source: This box draws on G. Berhane et al., *The Implementation of the Productive Safety Nets Programme, 2014: Highlands Outcomes Report (2015)* (Addis Ababa: Ethiopia Strategy Support Program, IFPRI, 2016); G. Berhane et al., "The Impact of Ethiopia's Productive Safety Net Programme on the Nutritional Status of Children: 2008-2012," ESSP Working Paper 99 (IFPRI and Ethiopian Development Research Institute, Washington, DC, and Addis Ababa, 2017); and J. Hoddinott et al., "The Impact of Ethiopia's Productive Safety Net Programme and Related Transfers on Agricultural Productivity," *Journal of African Economies* 21, no. 5 (2012): 761–786.

BOX 5 INDIA IMPLEMENTS NEW ASPIRATIONAL PROGRAM FOR DISTRICTS LEFT BEHIND

Purnima Menon (IFPRI)

The rate of childhood stunting has fallen over the past decade in India, from 48 to 38 percent. However, stunting and other nutrition, health, education, and economic outcomes vary widely across and within states, with tribal and remote areas being particularly vulnerable. In recognition of India's malnutrition challenges, the Indian government launched a National Nutrition Mission in early 2018, which is led, supported, and monitored nationally but financed and implemented by both national and state governments. Reflecting decentralization, the mission's strong district-level focus for nutrition reaches all districts in India. In the most vulnerable parts of the country, the mission is linked to another ambitious program, the Aspirational Districts Program.

The Aspirational Districts Program aims to change the narrative from "backward" districts to "aspirational" districts, with emphasis on using data and evidence to support and nudge districts to close gaps in implementation. By ranking districts on change in selected nutrition, health, education, and economic outcomes, the program aims to use competition and innovation to improve governance and program implementation in 112 districts.

What does this combined focus of the National Nutrition Mission and the Aspirational Districts Program mean for improving nutrition and inclusion outcomes? Initial observations suggest that the approach is indeed nudging districts to close gaps in governance, increase coverage of services, and generate greater interest in nutrition as a development issue. Given the multisectoral nature of the determinants of poor nutrition, a wide-scale, development-oriented effort like the Aspirational Districts Program has the potential to influence many of the known social determinants of poor nutrition outcomes–lack of education and health services, poverty, early marriage, and more–while also influencing the governance of core health and nutrition interventions. Although food system transformation is not a district-level mandate, districts will contribute to inclusive food system transformation through efforts to improve the functioning of India's public food programs.

Note: Drawn from work by the Partnerships and Opportunities to Strengthen and Harmonize Actions for Nutrition in India (POSHAN) program, with support from A4NH (see poshan.ifpri.info).

BOX 6 MALAWI ADOPTS A FOOD-SYSTEMS APPROACH TO IMPROVING DIETS AT THE COMMUNITY LEVEL

Aulo Gelli (IFPRI)

Two recent impact evaluations in Malawi show how interventions at the community level can improve diet quality for the poor. In the first, findings suggest that during the lean season in food-insecure settings, where people face declines in food security, diet quality, and nutrition status, food transfers can have a protective effect on diets of low-income populations. Food transfers may also create demand that supports markets for nutritious foods. An evaluation of these food transfers found a 15 percent increase in children's dietary diversity scores, including foods not involved in the transfer, like vegetables and dairy. This suggests households used the increased resources to prioritize the consumption of nutritious foods. In a second evaluation, an integrated agriculture and nutrition intervention was shown, after 12 months, to increase nutritious food production, production diversity, and maternal knowledge, and to improve nutrition practices at the household level, the diets of preschoolers, and linear growth in their younger siblings.

This evidence highlights the potential to boost the dietary impact of Malawi's existing social protection interventions by enhancing public and private sector linkages across the food system. During the lean season, the effectiveness of food transfers could be maximized by systematically integrating intensive behavior change communication (BCC) to optimize household food choices, and public procurement programs like the existing school meals program could be modified to purchase leafy green vegetables, increasing demand in village markets where these foods are highly available but buyers are few. In the postharvest period, public procurement could continue, accompanied by BCC to improve food choices, thus providing a steady demand for food system transformation.

Source: Adapted from A. Gelli et al., "Value Chains to Improve Diets: Diagnostics to Support Intervention Design in Malawi," *Global Food Security* (forthcoming), https://doi.org/10.1016/j.gfs.2019.09.006; and A. Gelli et al., "Using a Community-Based Early Childhood Development Center as a Platform to Promote Production and Consumption Diversity Increases Children's Dietary Intake and Reduces Stunting in Malawi: A Cluster-Randomized Trial," *Journal of Nutrition* 148, no. 10 (2018): 1587–1597.

NEW OPPORTUNITIES AND CHALLENGES

National food systems are an important entry point for improving sustainability, health, and equity outcomes, and food system transformation is considered a key pillar of development in LMICs. Taking a food systems approach allows countries to consider a wide range of current challenges, from food security to climate change to diet transition, and opportunities, such as digital technology and building youth skills and entrepreneurial capacity, that are most relevant to their particular contexts. For LMICs, a food systems approach provides broad benefits in terms of designing effective, inclusive policies that can contribute to better nutrition outcomes. At present, a food systems approach is not proactively applied in most countries. This is reflected in policies that are supply-led rather than demand-led and which fail to anticipate food system transitions.

Inclusion and nutrition outcomes are closely linked. The burden of malnutrition is significantly greater among marginalized groups. National food system transformation strategies must be aggressively augmented with coordinated efforts to support groups left behind. In addition, we have reviewed some interesting examples of inclusion strategies at the district and community levels that can complement national food system transformation strategies. However, food system transformation strategies have not yet been systematically embraced in LMICs.[30] This is a critical moment for developing more systematic approaches to inclusive and healthy food systems—systems that encompass diet-led policies; build on synergies across technology, institutions, and infrastructure; and create an enabling policy environment to bolster the contribution of the private sector. Promoting national efforts to strengthen analysis and action can help to change the current trajectory in favor of healthier, more sustainable, and more equitable outcomes for all, for decades to come.

REGIONAL DEVELOPMENTS

BUILDING INCLUSIVE FOOD SYSTEMS TO HELP REACH THE GOAL OF ENDING HUNGER and malnutrition globally will require innovation and investment at the regional and country levels. This section discusses problems, policies, and prospects for regional and national food systems in 2020 and beyond across the major regions: Africa, the Middle East and North Africa, Central Asia, South Asia, East and Southeast Asia, and Latin America and the Caribbean. The challenges and potential of inclusion to transform food systems for better well-being and nutrition are examined for each region, along with other current topics:

- Africa's initiatives to create opportunities for women and young people in agricultural value chains
- Farm and agrifood system jobs for refugees and migrants in the Middle East and North Africa
- Impact of labor migration on household incomes and women's role in agriculture in Central Asia
- Efficiency of social transfer programs in South Asia
- African swine fever's impact on food production and consumption in East and Southeast Asia
- Obesity and overweight in Latin America and the Caribbean

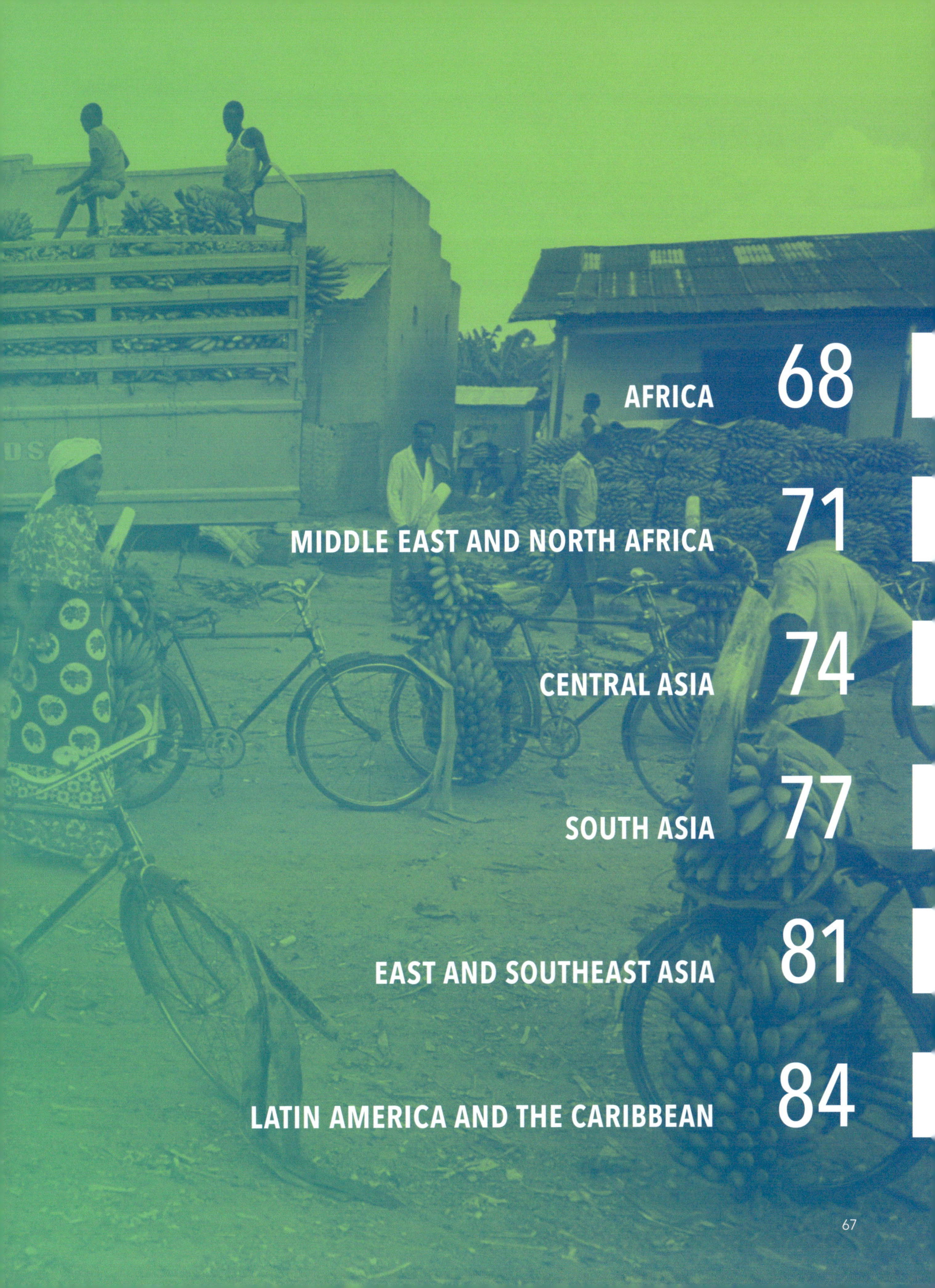

AFRICA

OUSMANE BADIANE, JULIA COLLINS, AND TSITSI MAKOMBE

Ousmane Badiane is the former director for Africa, International Food Policy Research Institute (IFPRI), Washington, DC, USA. **Julia Collins** is a senior research analyst, and **Tsitsi Makombe** is a senior program manager, Africa Region, IFPRI.

Income growth, economic dynamism, and demographic change in Africa are transforming food systems and changing opportunities for farmers, entrepreneurs, and employees along the agrifood value chain. Recent economic growth in most countries has created broadly shared benefits, including higher incomes, improved nutrition, and reductions in the prevalence of poverty and in the poverty gap (a measure of the severity of poverty among those who remain poor).[1] Income inequality, while high, has not increased over time. But poverty, hunger, and vulnerability are persistent, and growth has also led to perceptions of exclusion among regions, communities, and individuals benefiting less. Large farmers are best-placed to take advantage of the opportunities created by increased food demand from urban markets, the rapidly expanding food processing sector, and the modernization of distribution chains. Smallholders are at risk of exclusion from value chains if they are not able to meet the demands of high-value markets.

Similarly, efforts to increase trade integration in Africa may increase inequality if some benefit more than others. The new African Continental Free Trade Area, for which the associated agreement entered into force in May 2019, is expected to allow African countries to increase exports, better weather economic shocks, and improve food security.[2] However, increased market integration can also lead to geographic reallocation of production and other activities along the value chain, giving rise to winners and losers.

African leaders acknowledge the importance of ensuring that economic growth provides benefits for all. In the 2014 Malabo Declaration on Accelerated Agricultural Growth and Transformation for Shared Prosperity and Improved Livelihoods, leaders committed to enhancing the resilience of vulnerable groups and creating opportunities for women and youth in agricultural value chains.[3] For example, in its five-year national agriculture investment plan launched in 2018, Malawi put forward measures to (1) enhance the participation of women, youth, and other vulnerable groups in farmers' organizations, (2) sensitize rural households and service providers on gender relations and land tenure, and (3) promote sustainable natural resource management, including through training targeted at women and youth.[4] Although exclusion exists along multiple dimensions, significant efforts have been made to increase inclusion among three groups: the rural poor, youth, and women.

INCLUSION LAGS IN RURAL AREAS

Poverty is unevenly distributed, with much higher rates in rural than in urban areas in most countries.[5] Access to basic services is also much more limited in rural areas, with urban residents two to three times as likely to have access to basic sanitation services, drinking water, and electricity (Figure 1).

FIGURE 1 Access to services for rural and urban populations, Africa south of the Sahara (percent)

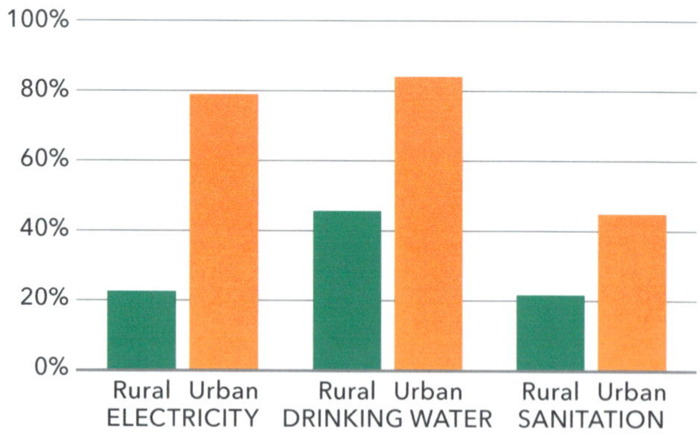

Source: World Bank, World Development Indicators database, 2019.

Note: Electricity = share of population with access to electricity; Drinking water = share of population using at least basic drinking water services; Sanitation = share of population using at least basic sanitation services.

Nutrition challenges too are usually more pronounced in rural areas. A recent study of nutrient adequacy in Senegal found that, while nutrient intake levels vary across the country, inadequacies tend to be more serious in rural areas. For example, urban residents in the majority of Senegal's departments have average vitamin A adequacy rates above 60 percent, while average adequacy rates are much lower in most rural areas (Figure 2). Addressing such imbalances will require closely monitoring seasonal and geographic differences in access to adequate nutrients and diverse diets in order to guide policies aimed at improving the affordability of foods, in both rural and urban markets, that meet standards for nutrient adequacy and dietary diversity.[6]

Increased investments in rural infrastructure and social services are important to reach vulnerable populations. Isolation and remoteness from services is strongly associated with poverty.[7] Improved access to transportation infrastructure and healthcare has been found to protect child growth from the effects of rainfall and production deficits.[8] Closer rural-urban linkages can increase market and employment opportunities for rural residents.[9] Innovations in digital services can extend the reach of extension and financial services and strengthen the capacity of farmer organizations to link smallholders with value chains (see Chapter 2). Governments can create an enabling environment for technological innovation by implementing appropriate regulation, providing incentives for private sector innovation, and investing in agricultural R&D and skills development.[10]

EMPLOYMENT FOR A GROWING YOUTH POPULATION

Africa's large youth population has great potential to contribute to economic growth and innovation (see Chapter 3). However, labor markets must create more productive and attractive employment opportunities for young workers. Unemployment among African youths aged 15 to 24 is estimated at over twice the overall unemployment rate.[11] High youth unemployment reflects mismatches in skills supply and demand. For example, only 2 percent of university students in Africa south of the Sahara are studying agriculture, despite the high share of the labor force engaged in that sector.[12]

Innovations in digital services, mechanization, processing, transport, distribution, and marketing present attractive employment and entrepreneurship opportunities for youth. However, skills gaps are a major constraint. Recent World Bank Enterprise surveys in seven African countries found that over 30 percent of

FIGURE 2 Household adequacy in vitamin A in Senegal, urban and rural areas (percent)

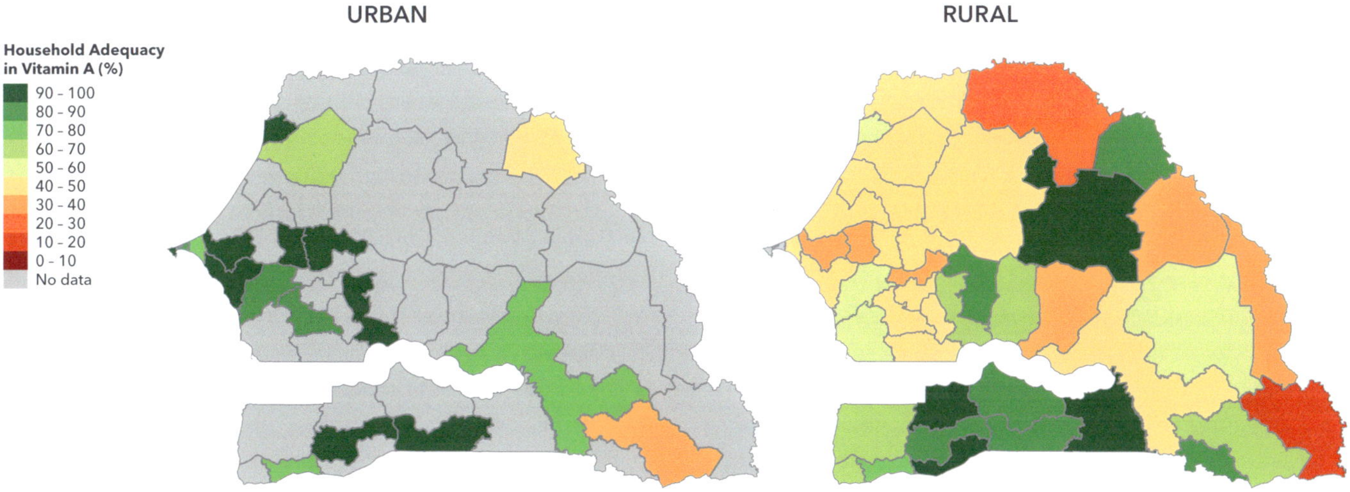

Source: J. Ulimwengu et al., "Hidden Hunger: Understanding Dietary Adequacy in Urban and Rural Food Consumption in Senegal," in *Transformation Structurelle du Système Agricole et Alimentaire du Sénégal*, eds. O. Badiane, M. D. Faye, and K. Savadogo (forthcoming).

surveyed firms in each size category, from micro to large, reported skills as the most severe constraint to their business operations.[13] Agricultural technical and vocational education and training systems are underfunded and underprovided in many African countries, and strengthening them should be a priority to provide not only youth but also older people with the skills required for emerging opportunities.

ADDRESSING GENDER INEQUALITIES THROUGH INCLUSION OF WOMEN

The African Union designated 2010–2020 as the African Women's Decade, with the goal of advancing the implementation of international gender equality commitments,[14] and many African countries and regional economic communities have implemented policies and strategies intended to promote greater equality. For example, the Southern African Development Community (SADC), comprising 16 southern African countries, launched the SADC Protocol on Gender and Development in 2008 and updated it in 2015.[15] The protocol calls on member countries to ensure women's constitutional and legal rights, achieve equal representation of women in political and other decision-making spheres, and advance equality in education and employment, among other areas. In October 2019, the government of Burkina Faso, in collaboration with the African Union, launched an initiative to advance mechanization to better meet women's needs and contribute to their empowerment in agriculture.[16] However, despite these and other national, regional, and continental initiatives, gender inequalities persist in educational attainment, political decision-making power, and employment, as well as in decision-making power at the household level (see Chapter 4).[17]

These inequalities have grave consequences, both for individuals and for economies. For example, unaddressed gender productivity gaps in agriculture lower overall agricultural productivity. Many studies have found that women farmers have less access to productive inputs than men, and in some cases receive lower returns to inputs. Suggested responses include interventions to increase women's access to farm labor and other inputs and to facilitate women's participation in markets and producer organizations.[18] Women's participation along the agricultural value chain—and the returns to their participation—can be improved through policies and projects with explicit gender equality goals, including strengthening the capacities of women and women's groups.[19] Rwanda's Strategic Plan for Agriculture Transformation (2018–2024) includes interventions to build women's entrepreneurial skills through training in leadership, management, and farming as a business, mentoring on agribusiness incubation, and providing business development support to enable women to access suitable financial products.[20]

THE WAY FORWARD FOR MORE INCLUSIVE FOOD SYSTEMS

Rapid growth in incomes and urbanization are transforming African food systems and giving rise to new opportunities along the value chain. But greater effort is needed to ensure that the benefits of growth are broadly shared. Upgrading physical and social infrastructure will not only improve livelihoods today but also boost future capacity to create wealth and reduce vulnerability in rural areas. Social protection, skills development and training, and other interventions, particularly targeted toward women and youth, can help vulnerable groups contribute to and benefit from agricultural transformation. For example, Ethiopia's Productive Safety Net Program has helped to protect poor households from consumption crises, lower food insecurity, and minimize disincentives to agricultural production.[21] Finally, inclusiveness requires a better understanding of exclusion and vulnerability and their drivers. Recent studies have shown the potential for finely tuned subnational targeting of food security interventions when sufficient data are available.[22] African countries should prioritize generating evidence on inequalities and on the winners and losers of major policy changes such as the African Continental Free Trade Area. The 2020 second CAADP Biennial Review of progress toward the commitments of the Malabo Declaration offers an opportunity to assess countries' success in tracking progress and in increasing the inclusion of smallholders, youth, and women.

MIDDLE EAST AND NORTH AFRICA

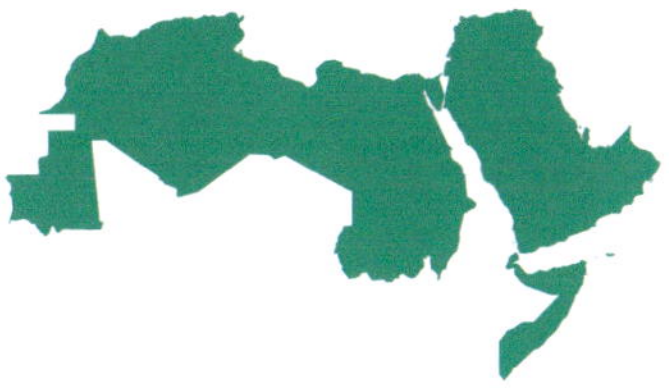

FATMA ABDELAZIZ, CLEMENS BREISINGER, NADIM KHOURI, AND JAMES THURLOW

Fatma Abdelaziz is a research associate in the Development Strategy and Governance Division, International Food Policy Research Institute (IFPRI), Cairo, Egypt. **Clemens Breisinger** is a senior research fellow in the Development Strategy and Governance Division and leader of the Egypt Strategy Support Program, IFPRI, Cairo, Egypt. **Nadim Khouri** is an independent researcher supporting the Global Agriculture and Food Security Program. **James Thurlow** is a senior research fellow in the Development Strategy and Governance Division, IFPRI, Washington, DC, USA.

Uncertainty was pervasive in the Middle East and North Africa (MENA) in 2019, reflecting ongoing conflicts and their regional spillover effects, global trade tensions, and fluctuating oil prices.[1] These problems continue to pose significant challenges, and as a result regional economic output (GDP) growth slowed in 2019 and is expected to have fallen to less than 1 percent.[2] Lower oil prices have disrupted oil exporters' plans to diversify away from oil because they now lack the funds to invest in new sectors.[3] Despite the lower prices, however, several oil-importing countries, including Egypt, Jordan, Lebanon, and Sudan, have exceeded the threshold for public debt that is considered sustainable for emerging markets.[4] This is limiting both public and private investments in these countries.

The private sector's role in delivering growth and employment in the region continues to be constrained by competition from public enterprises, red tape, a scarcity of skilled labor, and barriers to trade.[5] As a result, the private sector in MENA countries faces challenges in creating enough jobs for the millions of young people entering the workforce every year.[6] Efforts to address unemployment are underway in several countries that are adopting more business-friendly policies.[7] But MENA countries must also continue reforms to foster inclusive growth and create jobs, especially for youth and women. In fact, including more women and young people in the labor market can itself be a driver of growth.[8] In addition, MENA countries can learn from the successful example of Egypt as to how social protection programs—and especially well-targeted cash transfer programs—can support the poorest when these households face economically challenging times.[9] A question raised in many MENA countries is whether agriculture and the broader agrifood system can play a consequential role in fueling economic growth, job creation, and inclusion.

FOOD SYSTEMS CAN SUPPORT INCLUSIVE ECONOMIC GROWTH

The MENA region is characterized by its aridity and crisis-level water scarcity in many countries. Climate change is exacerbating this situation. This means that, far from abandoning the agriculture sector, countries should provide strategic direction for the sector to ensure more efficient and productive use of water.[10] While agriculture continues to be an important economic sector in most MENA economies, it is even more so when the entire agrifood system is considered (Figure 1). In countries such as Egypt, Jordan, Morocco, and Tunisia, the share of agriculture (farming) in GDP is between 7 and 15 percent, and the share of agriculture in employment is generally higher, at between 4 and 39 percent.[11] When all upstream and downstream agriculture-related sectors (trading, processing, food services) are considered, the agrifood systems in those four countries account for 19 to 27 percent of GDP and 21 to 45 percent of employment. Global data suggest that, among lower-middle-income countries, the contribution to GDP from the off-farm components of the agrifood system is greater than that of farming. And among high-income countries, more agrifood system jobs exist off the farm than on the farm. Recognizing this evolution of agrifood systems will be critical for development strategy and planning processes as MENA countries become wealthier.

FIGURE 1 Share of the agrifood system in GDP and employment

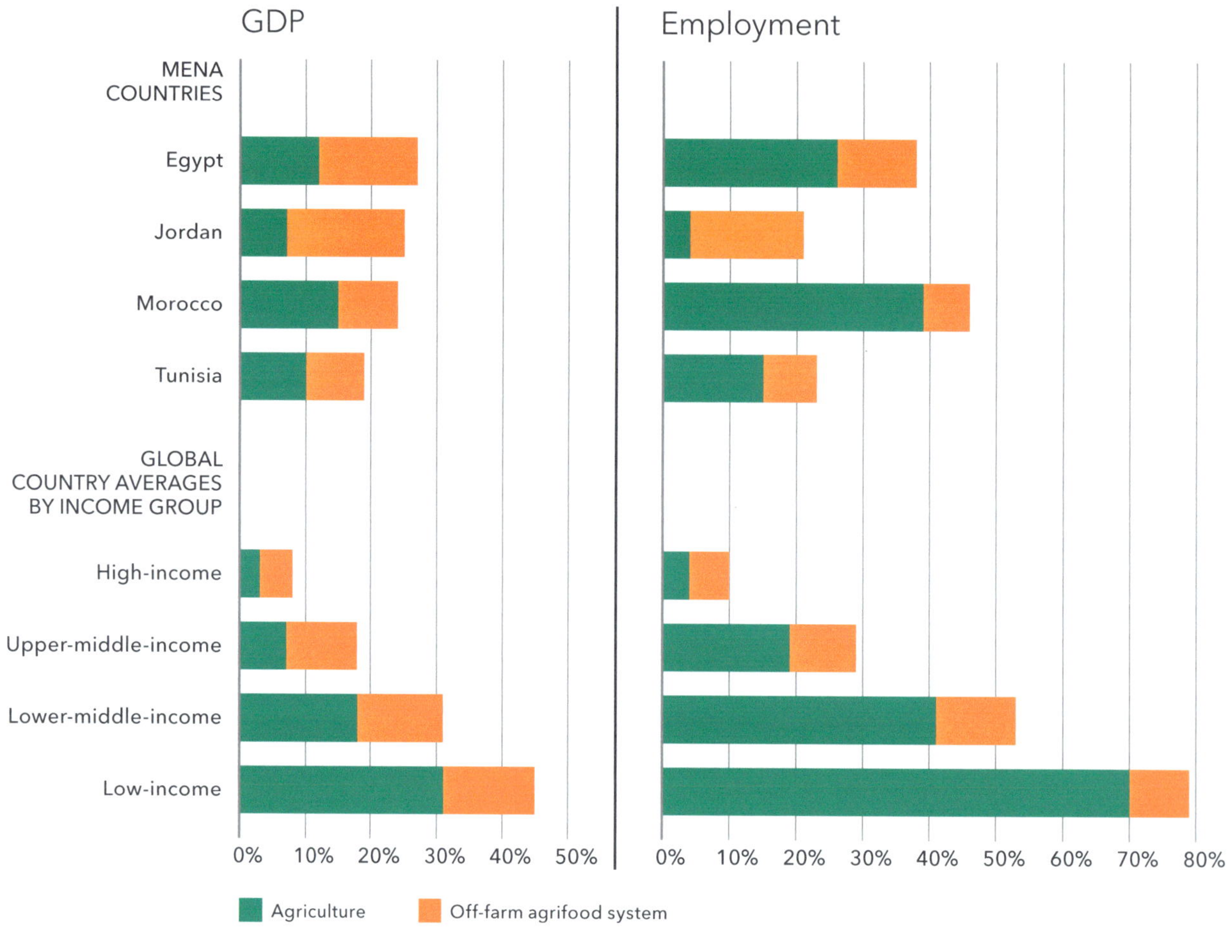

Source: J. Thurlow, "Measuring Agricultural Transformation" (PowerPoint presentation to USAID, Washington, DC, 2020), https://www.slideshare.net/ifpri/aggdp-agemp-measuring-agricultural-transformation.

Taking a holistic agrifood systems approach helps to better identify the full impact of agriculture sector development. In Lebanon, for example, the agriculture sector has been stagnating, with growth at roughly zero, while the food processing sector, a key component of the country's agrifood system, grew at 5 percent annually from 2005 to 2010, followed by slower but still respectable growth of 2.5 percent annually between 2010 and 2015; Lebanon's agrifood sector has been identified as one of five sectors with the strongest potential to promote the country's economic aspirations.[12] While the agriculture sectors in Egypt and Morocco continue to grow, labor has moved from farm to nonfarm sectors as people seek to overcome the low productivity trap and the informal nature of employment in agriculture.[13]

Indeed, more broadly in MENA and consistent with global transformation trends, the rate of increase in farming jobs was one of the lowest, at 5 percent between 2010 and 2016, compared with other employment opportunities that have seen significant increases, such as start-ups in food services, technician positions in food processing, and jobs in quality control.[14] In response to this transformation, some MENA countries have begun shifting their policies and investments away from primary production toward value-added sectors of the food economy. For example, in 2019, the United Arab Emirates announced a US$272 million incentive package for agritech, and Saudi Arabia's Agriculture and Livestock Investment Company continues to invest in the food processing sectors of MENA countries.[15]

TOWARD COMPREHENSIVE AGRIFOOD SYSTEM STRATEGIES AND INVESTMENTS

The region and the global community urgently need to resolve MENA's protracted conflicts and to address the pressing needs of refugees, internally displaced people, and those living in conflict zones, especially women and children. Food systems offer not only a means to provide emergency assistance to those in need–including in conflict and crisis situations–but also ways to reduce the potential for food insecurity to contribute to conflict. Food systems also offer an obvious starting point to promote economywide growth and employment during reconstruction and recovery in countries like Yemen (see Chapter 5).[16] Many jobs in farming and the agrifood system are already done by vulnerable groups such as refugees, migrants (for example, in Lebanon and Jordan), and the poor, so a food-system-led transformation is also likely to foster *inclusive* transformation for these groups.

Increasing incentives for the private sector–on its own or through private-public partnerships–to invest in all segments of agrifood systems can be one important way to promote such a transformation. Despite countervailing trends observed in some countries, support for investments in the agriculture and agrifood sectors tends to be more beneficial than protectionist policies, such as import tariffs and subsidies, in terms of promoting sector growth and food security.[17] However, it is important to note that ongoing global trade tensions and related uncertainties strengthen the position of policymakers who argue for increasing food self-sufficiency and challenge the advice of most economists that countries should focus domestic food production in areas of comparative advantage (for MENA, that broadly means exporting fruits and vegetables and importing cereals). In addition, the experience of Jordan and Lebanon–which lost much of their traditional food export markets in Iraq and Syria due to conflict–highlights the risks for countries following an export-led food strategy and points to the importance of diversifying their export markets.

For large MENA countries, and especially for smallholders within those countries, fostering growth of domestic markets can be more supportive of inclusion than growing export markets. Expanding domestic markets can give smallholders more time to reach the necessary quality standards for exports. An inclusive, food-system-led transformation will also benefit from the adoption of institutional innovations (such as establishing presidential delivery units and transformation agencies) as well as innovations that help cope with intensifying challenges such as water scarcity and climate change.

To go beyond such general advice, evidence-based and country-led food policy and investment analysis is needed. Digitizing food policy analysis and using "big data" are critical steps in this direction. An example of such an effort is the Agricultural Investment Data Analyzer (AIDA, a joint project of IFPRI, the International Fund for Agricultural Development, and the CGIAR Research Program on Policies, Institutions, and Markets), which allows for prioritizing agricultural investments by "modeling without a model" through an online interface.

More generally, improvements in the region's education systems should prepare people for the gamut of employment opportunities available now and in the future in food systems and in food-related research, innovations, and policymaking. More effectively including women and youth in this process and at all levels will not only improve their personal well-being and livelihoods but will also greatly contribute to economic growth and transformation.

CENTRAL ASIA

KAMILJON AKRAMOV, ROMAN ROMASHKIN, ALLEN PARK, AND JARILKASIN ILYASOV

Kamiljon Akramov is a senior research fellow in the Development Strategy and Governance Division, International Food Policy Research Institute (IFPRI), Washington, DC, USA. **Roman Romashkin** is the head of the Department of Economics of the Eurasian Center for Food Security, Lomonosov Moscow State University, Moscow, Russia. **Allen Park** and **Jarilkasin Ilyasov** are senior research analysts in the Development Strategy and Governance Division, IFPRI.

In Central Asia, addressing poverty and unemployment among women and youth is essential to creating inclusive food systems. The share of working-age people in the region's population has been gradually increasing over several decades (Figure 1), and today a large cohort of young people and women cannot find employment and earn adequate income in their own countries. For example, in 2019, the unemployment rate for women in Uzbekistan stood at 12.8 percent, and the unemployment rate for youth (between the ages of 20 and 30) stood at 15 percent.[1] In Central Asia's rural areas, rates of unemployment for women and youth are significantly higher, which is reflected in high rural poverty rates. In the rural mountainous region of Naryn Province in Kyrgyzstan, for instance, youth (ages 15 to 29) unemployment stood at 22 percent in 2018, and for young women it was above 40 percent.[2] Similar unemployment rates for youth and women are observed in rural areas of other Central Asian countries. Across the region, youth (ages 15 to 29) currently make up about 25 to 30 percent of the population, and this share is expected to remain high for the foreseeable future.

FIGURE 1 Working-age population (20–59 years old) in Central Asian countries, as share of total population

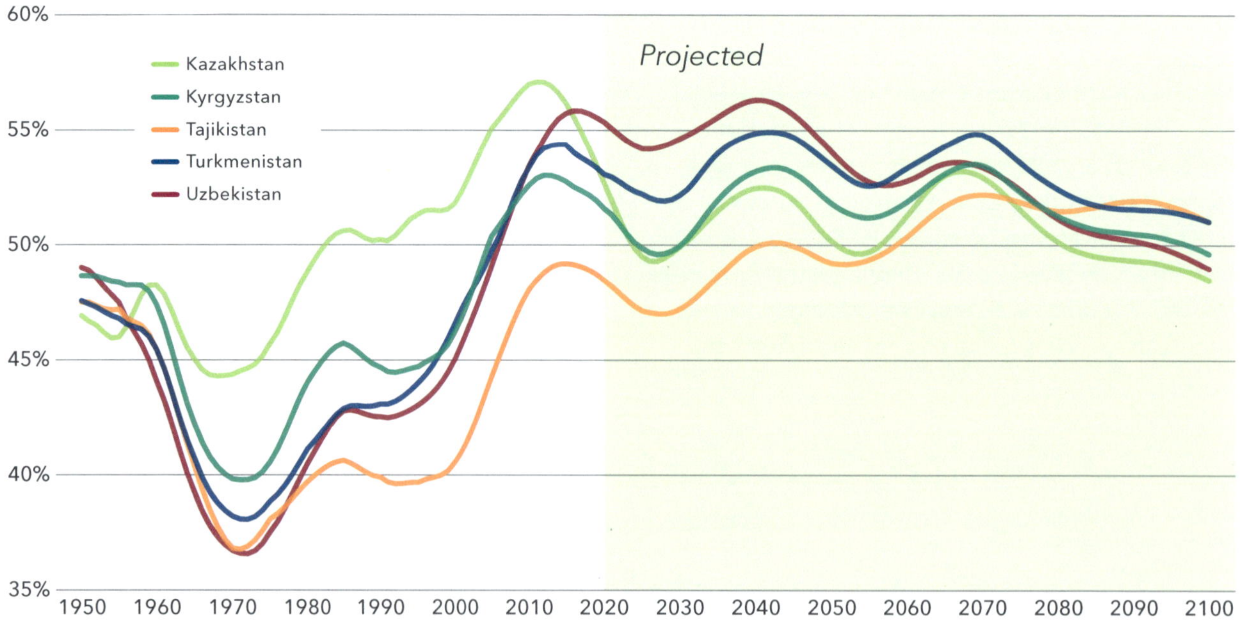

Source: Based on United Nations, Department of Economic and Social Affairs, Population Division, "World Population Prospects, 2019," accessed January 2, 2020, https://population.un.org/wpp/Download/Standard/Population/.

EMPLOYMENT, MIGRATION, AND INCLUSION

Creating employment opportunities for youth and especially for young women will be essential to improving the inclusiveness of food systems, including increasing the income, equity, and nutrition benefits of food systems for rural people. Promoting high-value agrifood sectors, such as horticulture, livestock, food processing,[3] and business activities along related value chains, such as logistics and storage infrastructure, may help to create employment opportunities, particularly in densely populated rural areas. Development of the horticulture sector has added benefits, including a significant positive impact on food access and nutrition outcomes—empirical evidence suggests positive linkages between crop diversity and dietary diversity in Tajikistan and between crop diversity and agricultural productivity in Kyrgyzstan.[4]

The lack of jobs at home forces Central Asian workers, especially young men, to seek employment in Russia, Kazakhstan, Turkey, and elsewhere. Migration from labor-abundant countries (Kyrgyzstan, Tajikistan, and Uzbekistan) became a major socioeconomic phenomenon over the past two decades. Labor remittances, which are slowly rebounding from the low levels experienced in 2015 and 2016 (Figure 2), are a critical source of foreign exchange in these countries. The inflow of remittances, primarily from Russia, contributes to macroeconomic stability, increased incomes, poverty reduction, and macro- and household-level food security in the region. However, evidence suggests that remittances tend to support consumption rather than providing capital for economic development, and can therefore have some unintended and negative consequences for structural transformation of remittance-receiving economies.[5]

Labor migration has two notable impacts on the inclusiveness of Central Asia's food systems. First, labor remittances can improve household welfare and access to food. For example, according to a recent IFPRI survey, about 40 percent of households in Tajikistan have at least one family member working abroad (usually in Russia) and receive remittances. For families that receive remittances, food amounts to about 50 percent of their expenditures, whereas for households not receiving remittances, nearly 55 percent of their expenditures are on food.[6]

FIGURE 2 Total remittance inflows from Russia (2010–2019, quarters 1–3)

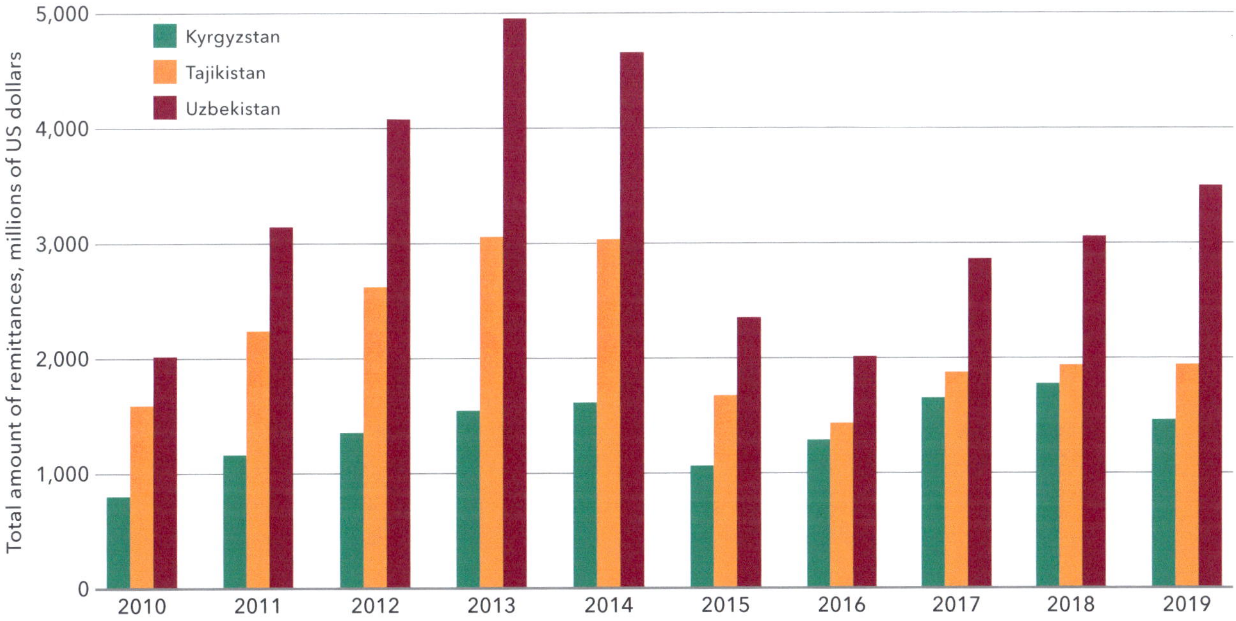

Source: Based on data from the Central Bank of the Russian Federation, "Cross-Border Transfers of Individuals, 2019," accessed January 2, 2020, https://www.cbr.ru/eng/statistics/macro_itm/svs/.

Second, labor migration from Central Asia tends to be predominantly male and rural, which leads to the "feminization" of agricultural labor (see Chapter 4). This can have both positive and negative outcomes: while earnings from remittances and increases in women's decision-making power can improve rural economies, men's migration can also contribute to agricultural labor shortages and create social issues.[7] This trend also highlights some of the institutional challenges related to inclusion in Central Asia's rural areas. For example, *dehkan*[8] farms in Tajikistan headed by women are often unable to access male-led water-users' associations, and consequently miss out on economic opportunities.[9] Some projects, such as the FAO's Promoting Inclusive Economic Growth Through Matching Grants initiative in Tajikistan, make a direct link between labor migration and agriculture in the migrant-sending countries by mobilizing the earnings and skills that migrants gained overseas for use in local agriculture and agribusiness.[10]

LOOKING FORWARD

Central Asia will continue to face global and regional risks related to climate change and commodity price uncertainties in the medium term. In addition, external vulnerabilities associated with political, economic, and trade conditions in the region's main trading partners (Russia and China) will have significant impacts on economic growth prospects as well as food and nutrition security. Because most of the region's agrifood sector exports are currently sent to Russia, diversification of export markets will be essential to improving the stability of Central Asia's economies and the development of food systems in the region.

The region's largest country, Uzbekistan, is considering joining the Eurasian Economic Union (EAEU) and applying for WTO membership in the near future. These steps will have important implications for transforming food systems in Uzbekistan and throughout the region. WTO membership could help harmonize national legislation and standards with international practices, increase predictability and transparency of the trade regime, and improve Uzbekistan's business and investment climate. It may also ease trade conditions with Uzbekistan's neighbors, which are already WTO members.

Membership in the EAEU would entail both risks and opportunities. First, it could improve employment opportunities in the Russian labor market for Uzbekistan's migrants, as they would not need to obtain and pay for work permits and other employment-related certification. Labor remittances could increase by up to 20 percent, and more than 2 million migrant laborers and their families (about 30 percent of Uzbekistan's population) could potentially benefit from these changes. In addition, the EAEU already accounts for about 30 percent of Uzbekistan's international trade. Uzbekistan's accession could create additional trade opportunities with EAEU members by harmonizing tariffs, removing customs controls at the borders with EAEU member countries, unifying transport and logistics regulations, strengthening coordination in the implementation of sanitary and phytosanitary measures, and synchronizing regional digital connectivity initiatives, including traceability of products and technology transfers. However, there is a risk that accession could lead to trade diversion effects by redirecting Uzbekistan's trade with non-EAEU countries toward EAEU markets and reducing the competitiveness of its exports in non-EAEU markets. Thus a careful assessment of potential impacts of membership in the EAEU and WTO on trade, household welfare, and economic growth would be necessary.

Kazakhstan and Kyrgyzstan recently updated their national development strategies, and Uzbekistan adopted a new agrifood sector development strategy for 2020–2030. These policy documents aim to transform food systems, promote nutrition-sensitive value chains, encourage private incentives and investments in the agrifood sector, and extend employment opportunities, especially for women and youth. The successful implementation of these strategies and policies requires the establishment of rigorous ex ante and ex post impact assessment frameworks, which will help identify policy and institutional constraints down the road and develop evidence-based policy solutions for promoting inclusive food systems in the region.

SOUTH ASIA

SHAHIDUR RASHID, AKHTER AHMED, AND ABDUL WAJID RANA

Shahidur Rashid is the director for South Asia, International Food Policy Research Institute (IFPRI), New Delhi, India. **Akhter Ahmed** is a senior research fellow and chief of party, Bangladesh Policy Research and Strategy Support Program, IFPRI, Dhaka, Bangladesh. **Abdul Wajid Rana** is the program leader, Pakistan Strategy Support Program, IFPRI, Islamabad, Pakistan.

South Asia's steady progress toward economic transformation has reshaped the region's diverse food systems over the past decade. This regional transformation has been marked by strong economic growth, rising real wages, and the expansion of nonagricultural sectors. The share of agriculture in national GDP has, on average, declined by 15 percent, and the share of nonfarm employment has now surpassed that of farm employment.[1] As these structural changes continue, the policy challenge lies in ensuring that food system transformation is inclusive and sustainable.

OUTLOOK FOR SOUTH ASIAN FOOD SYSTEMS

The Green Revolution led to remarkable growth in yields and overall output of cereal production in South Asia over the past five decades. But in recent years, the growth rate of high-value foods has been greater than that of cereals. The gross value of production (at 2004/05 constant prices) of high-value products—that is, milk and milk products, meat, and fruits and vegetables—grew by over 4 percent between 2000 and 2010 and by about 3 percent from 2011 to 2017, compared with 2.3 percent and 0.5 percent for cereals in these time periods, respectively. The magnitude of these changes varies by country. For instance, between 2010 and 2018, the value of meat production grew by about 7 percent in Bangladesh and Bhutan and 5 percent in Pakistan, but by less than 1 percent in Sri Lanka and less than 2 percent in India and Nepal.[2]

Changes in food production are mirrored in the region's food consumption. Cereals are rapidly losing their importance in household food baskets, particularly among poor households (Figure 1). As cereal consumption has declined, per capita consumption of

FIGURE 1 Cereal consumption in South Asia (as percent of food expenditure)

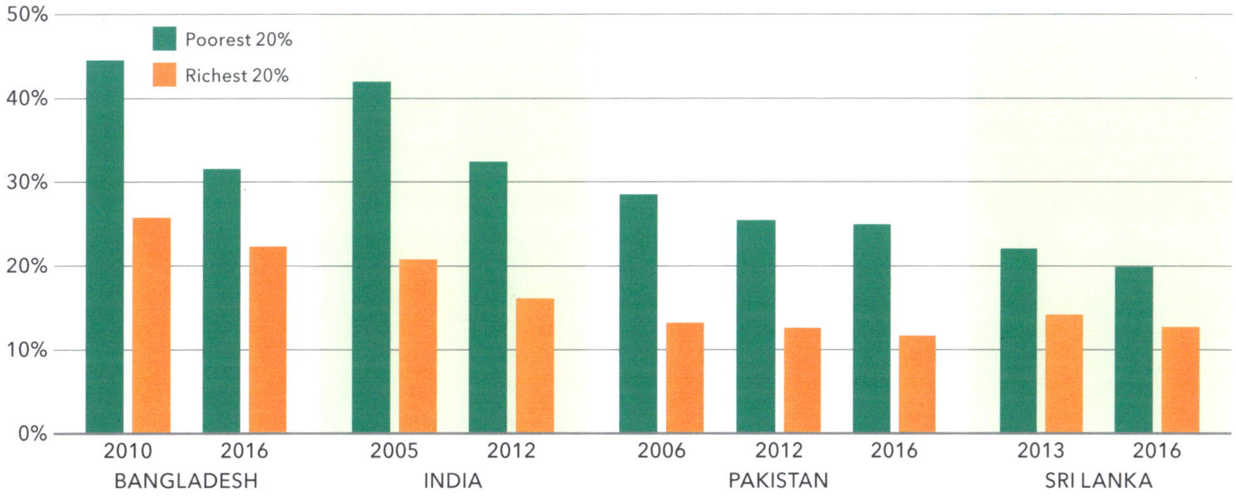

Source: Authors' calculation based on Household Income and Expenditure Survey in Bangladesh (2011 and 2019); National Sample Survey in India (2006, 2007, 2013, and 2014); Household Integrated Economic Survey in Pakistan (2007, 2013, and 2017); and Household Income and Expenditure Survey in Sri Lanka (2018).

meat, eggs, and fish has increased by over 40 percent; fruits and vegetables by 24 percent; and milk by over 10 percent.[3] In Bangladesh, the consumption of more diverse diets has contributed to measurable impacts on nutrition, including significant reductions in child stunting (from 43 percent in 2007 to 31 percent in 2017), underweight (from 41 percent to 22 percent), and wasting (from 17 percent to 8 percent).[4]

In line with income growth and demand for greater diet diversity, the food processing sector is also growing. Gross value added from food processing in India jumped from $6.9 billion in 2006 to over $16 billion in 2017. Similarly, gross value added more than doubled in Pakistan between 2000 and 2006 (from $1.3 billion to $3.4 billion).[5] At the regional level, food and beverage processing as a percentage of value added in manufacturing is estimated at 14.4 percent, with the share reaching 36 percent in Sri Lanka.[6] Yet postharvest losses are estimated to be higher in South and Southeast Asia than in other regions.[7] This suggests that the region could benefit from upgrading postharvest technologies, which in turn would increase food availability and contribute to environmental sustainability.

Data and robust studies on the impact of this food system transformation on inclusiveness are limited, but existing studies point to positive impacts for the poor. First, studies suggest that the poor are benefiting from new value chains—such as poultry and fisheries—that are emerging to meet changing consumer demand. Expansion of aquaculture in Bangladesh has contributed to job creation, poverty reduction, and better diets.[8] Similarly, promotion of poultry has proved to be pro-poor in South Asian countries.[9] Second, the growing food processing sector is generating employment for the poor. The number of jobs in food processing industries jumped from 1.4 million in 2006 to over 1.8 million in 2017 in India, and from 0.8 million to 1.7 million in Pakistan. In Bangladesh, more than 0.3 million jobs were created in the sector in 2012.[10] Finally, real agricultural wages are rising in almost all countries in the region (Figure 2). This is a remarkable success, especially given that the region has a very large rural labor force and real wages remained stagnant for decades, even following the Green Revolution.[11]

POLICY LEVERS FOR INCLUSIVE FOOD SYSTEMS

Three policy levers will be critical in making food system transformation inclusive and sustainable: (1) reforming agricultural input subsidies and price supports; (2) improving the targeting of social protection programs; and (3) building effective institutions for governing the emerging food system.

FIGURE 2 Real wages for agricultural workers in South Asia

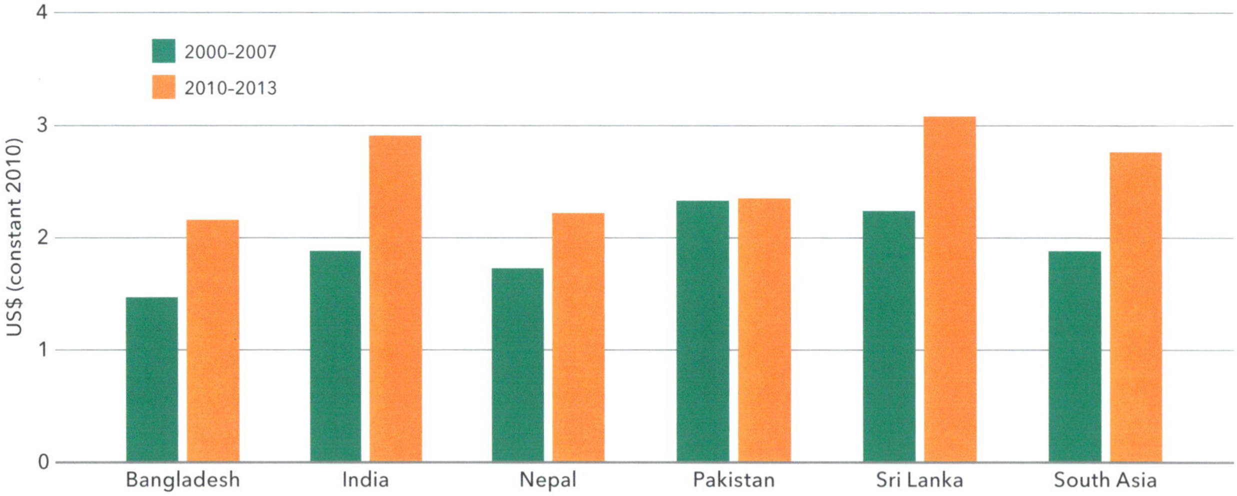

Source: Adapted from S. Wiggins and S. Keats, *Rural Wages in Asia*, Overseas Development Institute report (London: ODI, 2014); ILO (International Labour Organization), ILOSTAT database, accessed November 2019.

The agricultural subsidy and price policies adopted decades ago across the region to promote the Green Revolution have become an integral part of the food system and are politically popular, but are well documented to be inefficient, distortionary, and inequitable.[12] Reforming these programs could free up public funds to invest in fostering more inclusive, equitable, and gender- and nutrition-sensitive food systems.

Social safety net programs are effective policy vehicles for making the food system transformation inclusive (see Chapters 2 and 6). South Asia already has an extensive system of food-based safety net programs and public food distribution systems. However, data suggest that coverage of social protection varies widely by country (Figure 3).[13] While over 90 percent of both poor and rich are covered in India (due to almost universal coverage of public distribution and other transfer programs), only about 4 percent of the poorest Bhutanese are covered by social safety net programs. Similarly, an evaluation of Bangladesh's largest social safety net program for rural destitute women, the Vulnerable Group Development program, found that only 43 percent of its recipients came from the poorest quintile, largely because the beneficiary selection criteria were not observable, verifiable, or strongly linked with poverty.[14] The large share of rich households benefiting from these programs in some countries, notably India, indicates that better targeting social protection programs and redesigning food-based programs could make them more efficient, more effective in reaching the poor with healthy and nutritious food, and even reduce intimate partner violence. Evidence from Bangladesh shows that women receiving transfers in conjunction with nutrition behavior change communication experienced 26 percent less intimate partner violence 6 to 10 months after the intervention ended and sustained reductions up to four years afterward.[15]

Establishing effective institutions for food system governance is another strong policy lever for inclusive food systems. Cohesive institutional frameworks to govern South Asia's food systems are just beginning to emerge. Food safety and standards authorities have been set up only recently—in India in 2011, in Bangladesh in 2015, and in Pakistan in 2017—and are not yet effective. For example, in examining Bangladesh's dairy sector, recent studies have found that virtually all of Bangladesh's milk supply is contaminated with antibiotics, detergents, and heavy metals as a result of contaminated feed, which could lead to antibiotic-resistant infections, cancer, and other serious noncommunicable diseases.[16] Similar stories of food safety problems make headlines in other countries as well, exemplifying the potentially egregious health consequences that emerge from

FIGURE 3 Coverage of social safety net programs in South Asia

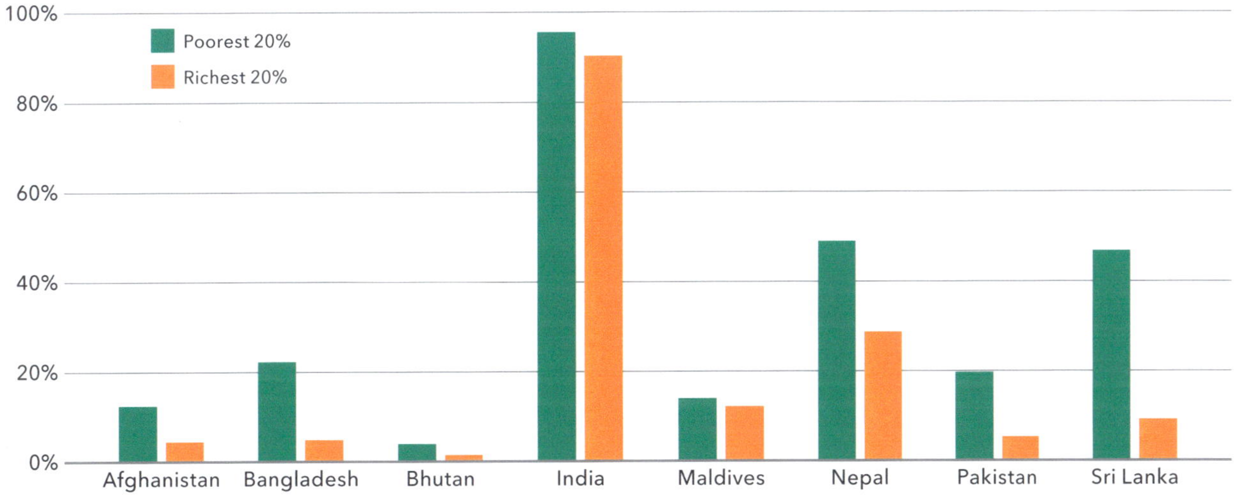

Source: Data from World Bank, World Development Indicators database, 2019.

food safety challenges.[17] Effective functioning of the new regulatory bodies will be critical to ensure food safety, and can also contribute to improving human well-being, creating market opportunities, and enhancing income through promotion of new agri-food value chains.

OPPORTUNITIES AND CHALLENGES FOR FOOD SYSTEMS

South Asia has made remarkable progress but faces new challenges and opportunities in making food systems inclusive and sustainable. The three policy levers discussed can create a range of opportunities. First, cutting the link between subsidies and cereals (rice and wheat) production will also incentivize farmers to diversify their crops, benefiting their incomes and dietary diversity. Second, better targeting of social protection programs, for example, using information technology to fine-tune social transfers and subsidies—as with India's Direct Benefit Transfer program and Pakistan's targeted subsidy program—will improve program efficiency and free up funds for other uses. Similarly, digitalization of social transfers can make transfer programs more efficient and contribute to inclusion, as has been the case with Bangladesh's government-to-person payment systems or Pakistan's Kifalat, which ensures financial and digital inclusion of poor women.[18] In addition, incorporation of behavior change communication into social safety nets can promote healthier diets among the poorest.

Despite the opportunities in these areas, however, challenges remain. For example, although the Direct Benefit Transfer program is a step in the right direction for India, reforming subsidies and agricultural price policies to meet new needs will remain difficult due to the political unpopularity of such efforts. Similarly, cereal-based safety net programs remain popular in the region despite ample evidence of better alternatives. Finally, new institutions for food system governance have significant potential for promoting an inclusive food system, but will require the political will to provide adequate funding and to respond to bottom-up pressure from consumer rights groups and civil society organizations. These organizations are gaining momentum in the region and will be essential to ensuring that the unfolding food system transformation is inclusive and sustainable.

EAST AND SOUTHEAST ASIA

KEVIN CHEN, PETER TIMMER, DAVID DAWE, AND MENGYAO LI

Kevin Chen is a senior research fellow, International Food Policy Research Institute (IFPRI), Beijing, and a chair professor, Zhejiang University, Hangzhou, China. **Peter Timmer** is professor emeritus, Harvard University, Cambridge, MA, USA, and a nonresident fellow, Center for Global Development, Washington, DC, USA. **David Dawe** is a senior economist and regional strategy and policy advisor, Food and Agriculture Organization of the United Nations, Bangkok, Thailand. **Mengyao Li** is a research analyst, Zhejiang University–IFPRI Center for International Development Studies, Hangzhou, China.

The year 2019 in East and Southeast Asia was marked by increasing uncertainty for the economy as a whole, and also for the development of inclusive and sustainable food systems. Although regional economic growth is expected to remain positive, the agricultural economies of the region face challenging prospects with the spread of African swine fever, weakening global demand for the region's exports, broadened trade disputes, and the outbreak of coronavirus.[1] Among the impacts on millions of residents, the livelihoods, food security, and nutritional status of vulnerable groups require particular attention.

TRANSBOUNDARY DISEASES THREATEN LIVELIHOODS AND FOOD SECURITY

The destructive fall armyworm is spreading in key corn-growing regions of East and Southeast Asia.[2] The pest poses a major threat to the region's corn farmers, many of whom rely heavily on sales of feed grain for household income. A decline in corn production will also force hog, poultry, and cattle growers to turn to more expensive feed imports.[3]

African swine fever is expected to have a noticeable impact on meat and feed markets worldwide. First reported in northeastern China in August 2018, the highly contagious, often fatal pig disease has reached eight countries across the region.[4] Despite imports of pork and government release of frozen stocks, the tight pork supply drove pork prices up by 21.3 percent in China in the first three quarters of 2019.[5] Rising prices are likely to lead consumers to turn to other types of meat and reduce pork consumption. In countries that are heavy consumers of pork, including Viet Nam, China, and South Korea, the most vulnerable consumers may suffer due to decreased diet quality, requiring further actions to tackle pork shortages and to control the rising prices (Figure 1).

Smallholder farmers account for a significant proportion of pig production but have limited biosecurity options to address African swine fever. As a result, they are among the hardest hit. To help forestall the spread of the disease, the Philippines, Viet Nam, Cambodia, Lao PDR, and China have implemented controls on the movement of pigs and pork products from affected communities. However, in China these restrictions have resulted in a widening gap in pork prices between producing and consuming provinces.[6]

FIGURE 1 Per capita pork consumption in 2018

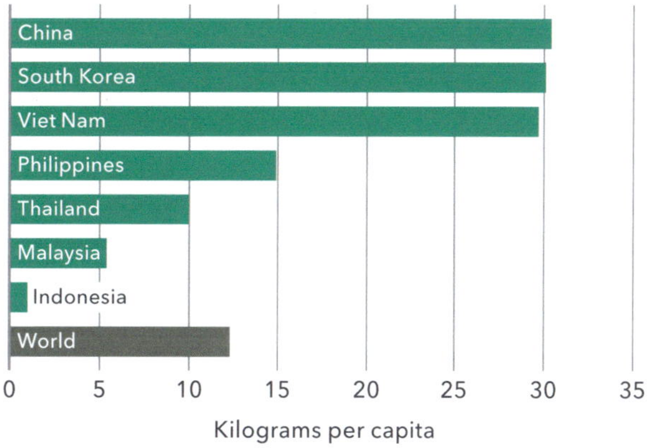

Source: Data from OECD/FAO, *OECD-FAO Agricultural Outlook 2019-2028* (Paris: OECD, 2019).

TRADE POLICY CHANGES AFFECT FARMERS AND CONSUMERS

Tariffs on agricultural products have been a weapon of choice as the US–China trade war has heated up, involving, at some points, more than 500 agricultural products including soybeans, grains, meat, dairy products, fruits, and nuts.[7] Chinese purchases of US agricultural products, including soybeans, are expected to increase under the "Phase 1" trade deal struck at the end of 2019, marking a step toward resolving the tit-for-tat tariff battle.

Against the backdrop of protectionism, East and Southeast Asian countries are working to deepen regional economic integration and advance the rules-based multilateral system, as indicated by the updated protocol for the China-ASEAN free trade agreement and the progress toward concluding the Regional Comprehensive Economic Partnership (RCEP) agreement between ASEAN and six other countries. ASEAN has benefited from several opportunities arising from the US-China trade turmoil and overtook the United States to become China's second-largest trading partner in the first half of 2019.[8] More tropical fruits, such as bananas and durian, are entering the Chinese market from Southeast Asia, providing new income-earning opportunities for farmers. China's palm oil imports from ASEAN countries are also surging as decreased US soy imports, in response to African swine fever and trade restrictions, have reduced Chinese production of soy oil and increased demand for other edible oils.[9]

Trade policies for rice have profound implications for the well-being of both producers and consumers in the region, as rice remains the major source of calories across Asia, especially for low-income families. In the Philippines, the removal of quantitative restrictions, long used to regulate rice imports and attain rice self-sufficiency, led to declining rice prices throughout 2019. As a result, per capita consumption of rice and daily calorie consumption are expected to improve, contributing to greater food security and better nutrition.[10]

However, Philippine smallholder rice farmers are struggling to stay profitable. For trade liberalization to be inclusive, it must be accompanied by appropriate interventions. The tariff revenue from rice imports has been earmarked for interventions intended to improve the competitiveness of the country's rice farmers (for example, support for mechanization and certified seeds). Targeted social safety nets (see Chapters 2 and 6), such as cash transfer programs to help smallholders cope with the price shock from increased rice supply, are another possible policy response. These complementary interventions are not likely to fully cushion the shock, however, and many farmers will need to change their cropping patterns, for example, by shifting to dry season vegetables.

DIETS ARE CHANGING, BUT NUTRITION CHALLENGES REMAIN

Despite continued economic growth, undernutrition remains a regional challenge and is widespread in the most vulnerable groups. The share of children under five who suffer from stunting (low height-for-age) averages 26 percent across ASEAN countries and constitutes a severe public health problem. The prevalence of wasting (low weight-for-height) and anemia also constitute moderate or severe public health problems. The prevalence of stunting is consistently highest in the lowest household wealth quintile. Furthermore, inequality is increasing, which slows the reduction of stunting in the poorest quintile for a given rate of economic growth.[11] Anemia particularly affects women of reproductive age and pregnant women in most countries.

A major contributor to undernutrition is the lack of dietary diversity in the region. In all low- and lower-middle-income ASEAN countries (except Viet Nam), the diets of more than half of all very young children (6–23 months) fail to meet minimum standards of diversity, leading to micronutrient deficiencies that affect child development and the potential of future generations.[12] A monotonous diet of rice and pulses provides 85 percent of daily calories but threatens nutrition improvement among Rohingya refugees from Myanmar, leading to notably worse nutrition outcomes among the displaced population compared with the host-country population. If countries are to end various forms of undernutrition, economic growth will need to be more inclusive so that a diverse diet providing nutrient adequacy is affordable to all.

At the same time, the risk of overweight and obesity is rising with rapid urbanization, as traditional diets are being replaced by foods higher in fats, salt, and animal products, usually with lower intake of fresh fruits and vegetables. All ASEAN countries are witnessing a rise in the number of obese adults.[13] China has the largest number of affected people worldwide, with about 46 percent of adults and 15 percent of children obese or overweight.[14] Diet-related noncommunicable diseases are top killers in the region and entail high social and economic costs for individuals, families, and countries.[15]

OUTLOOK FOR 2020

The year 2020 is likely to be the most difficult since 2009 for the region in several critical dimensions. Despite a reasonably stable regional rice market, African swine fever and substantial trade dislocations threaten the income and nutritional well-being of the region's most vulnerable households. These households are especially difficult to reach via state-sponsored social safety nets, so a return to a more stable food supply will be essential in 2020 to position countries of the region to address food security concerns. Adding to the instability is the outbreak of the new coronavirus (COVID-19 or NCP), which originated from a wild food wet market in Wuhan, China. The virus has spread quickly across China and around the world, causing the loss of thousands of lives and large economic losses, since its onset in December 2019. Various border controls both within China and at China's international borders have been introduced to contain the disease. While these controls may be necessary, they have disrupted food and nutrition security in China and beyond. As the number of people infected continued to rise in early 2020, it is clear that the impacts of the virus on food security must be monitored closely.

LATIN AMERICA AND THE CARIBBEAN

EUGENIO DÍAZ-BONILLA AND VALERIA PIÑEIRO

Eugenio Díaz-Bonilla is the head of the Latin America and Caribbean Program, International Food Policy Research Institute (IFPRI), Washington, DC, USA. **Valeria Piñeiro** is a senior research coordinator in the Markets, Trade, and Institutions Division, IFPRI.

A REGION IN TURMOIL

In Latin America and the Caribbean (LAC), governments are facing the impacts of low global commodity prices, worsening economic conditions, and related domestic turmoil, all of which have implications for the region's food systems and social inclusion. From 2001 to 2011, a sustained period of economic growth was supported by strong global demand for commodities ranging from oil and copper to soybeans and coffee, to name just a few. The downturn in commodity prices that followed slowed annual growth in per capita income to about 0.2 percent between 2012 and 2018. Economic contraction continued in 2019, with per capita income growth projected to have fallen to -0.9 percent.[1] Prospects for growth rebounding in 2020 remain very low.

Most countries have been affected by the regional downturn. The economic crisis in Argentina that began in 2018 deepened in 2019 and led to the incumbent president's defeat in the October elections. The new administration has increased export taxes on a variety of agricultural products to try to improve fiscal accounts, which could slow exports. However, the real exchange rate will likely remain at more competitive levels, which would counterbalance the disincentives associated with the export taxes and help maintain overall agricultural and food export levels.

The humanitarian and political crisis in Venezuela continues, with little prospect for resolution. The dire economic and social conditions continue to fuel a steady out-migration, particularly to Colombia. Altogether, the United Nations estimates that by the end of 2019, the total number of Venezuelan refugees since the crisis began will have reached four million (about 12 percent of the total population) (see Chapter 5).

The Mexican government inaugurated in December 2018 has made support of agriculture and of small and family farmers a priority. At the same time, migrants and asylum-seekers from Central America continue to flee insecurity and poverty, which are exacerbated by drought and crop failures. The resulting increase in arrivals at the US-Mexico border led to diplomatic disputes between the United States and all countries involved.

The new Brazilian government (inaugurated in January 2019) announced a more permissive stance on agricultural production and mining in the Amazon than that of past governments. This has generated concern worldwide, given the key role of the Amazon forest for global environmental sustainability, but the new government has asserted that decisions on managing these resources are an internal matter of Brazilian sovereignty.

Other countries in the region, including Chile, Ecuador, and Bolivia, were affected by strong social and political protests in 2019, several of which, at the time of this writing, were still evolving.

Several important trade developments have implications for regional and global agriculture. First, the free trade agreement between the United States, Mexico, and Canada, which is intended to replace the previous NAFTA agreement and was signed in October 2018, has been ratified by all members. Second, after more than 20 years of negotiations, Mercosur and the European Union announced in June 2019 that they have reached a comprehensive trade agreement. The specific details, however, seem to need additional work, and controversies about environmental issues in the Amazon have delayed further advances. Third, the US-China conflict may have helped some LAC countries, for example by boosting exports of some agricultural products from

Brazil and Argentina to China and of some industrial goods from Mexico and Central America to the United States. However, by adding further uncertainties to a weak global economy, the US-China conflict is also negatively affecting the region as a whole.[2]

MAKING FOOD SYSTEMS MORE INCLUSIVE

The process of urbanization, along with the expansion of infrastructure and the growth of intermediate cities (a topic discussed in IFPRI's *2017 Global Food Policy Report*), the greater integration of rural and urban markets, and the increase in foreign investment in processing, retail, and fast-food chains and restaurants, has led to the expansion and greater complexity of food systems in the region.[3]

Those food systems now represent important shares of value added and employment when we consider producers, inputs and services providers, traders, agro-industrial processors, a variety of retail outlets, and prepared-food providers–ranging from street vendors to formal restaurants.

Figure 1 shows LAC countries classified according to relevance of the food system in the economy, as measured by combining the percentages of agricultural and agro-industry value added and of rural employment. Bolivia, Ecuador, Honduras, Paraguay, and Peru show a higher dependence on the food system for value added and employment (a combined average of more than 20 percent), followed by Argentina, Colombia, Costa Rica, El Salvador, Jamaica, Mexico, Panama, and Uruguay with intermediate levels (about 10 to 20 percent). For the remaining countries, the indicator is below 10 percent.[4]

Given the scale of food system value chains in the region, they offer important possibilities for broad-based employment and inclusion for a variety of actors, including women, youth, and vulnerable ethnic groups.[5] A longstanding debate has considered whether small and family farms can integrate into, and benefit from, expanding value chains. Several studies suggest that small farms (not just large farms) can be beneficially integrated, but generally not the poorest ones.[6]

Female participation rates in primary production are lower in LAC than in other developing regions. Although most countries in the region have laws and programs intended to improve the inclusion of women, evidence clearly points to discrimination against women in access to land, credit, and technology, and in governance structures more generally (see Chapter 4).[7] Analysis of women's other roles in the food system, for example, as traders of fresh products, workers in agro-industries, and operators of a variety of food outlets, has been more limited.[8] It should be also noted that women (adolescents and adults) appear more affected by the worsening epidemic of overweight and obesity in the region, particularly in countries whose overall rates of overweight and obesity are especially high, like Chile and Mexico. Therefore, food systems need to be analyzed not only from the point of view of employment and inclusion of women, but also to better understand how women's empowerment in food systems can improve diets and reduce overweight and obesity. The need for data collection and evaluation of policy interventions is even more acute for understanding inclusion of youth and vulnerable ethnic groups in food systems.

FIGURE 1 Food system relevance in the economy

Source: Data from World Bank, World Development Indicators database, 2019.

The challenge of restructuring food systems for employment and inclusion is compounded not only by the fact that these food systems are a source of obesogenic diets, but also because they are a significant source of greenhouse gas emissions and other negative environmental externalities.[9] In fact, food systems are at the center of a variety of economic, social, environmental, and health outcomes, with implications for many of the Sustainable Development Goals.[10]

LOOKING AHEAD

Deteriorating economic and social conditions in LAC signal that food security and nutrition are likely to worsen in 2020. Food systems in LAC are already struggling with sustainability issues (ranging from deforestation to food waste and loss) and health challenges (in part associated with obesity). These systems should also be analyzed to understand their implications for employment and implementation of the substantive changes needed to place them on a stronger environmental and nutritional footing. An extensive policy research program on these topics could help to improve the overall functioning and inclusiveness of food systems in LAC and to achieve the SDGs by 2030.

"If we build on innovations and continue to pioneer new ideas, we can design food systems that are inclusive, climate smart, and sustainable, and we can provide healthy diets for everyone."

FOOD POLICY INDICATORS: TRACKING CHANGE

DECISION-MAKERS AND POLICY ANALYSTS NEED SOLID EVIDENCE AND TIMELY information to develop and implement effective food policies. The International Food Policy Research Institute develops and shares global public goods—including datasets, indicators, and indexes—as part of its mission to provide research-based policy solutions that sustainably reduce poverty and end hunger and malnutrition. This information can be used to gauge the impact of policy changes and the progress made on specific aspects of development.

This section highlights five of the indicator series generated by IFPRI research and illustrates some recent trends revealed by these datasets. Indicators include investments in agricultural research, public spending on agriculture, capacity for food policy research, agricultural total factor productivity, and projections for agricultural production, food consumption, and risk of hunger to 2030 and 2050. The full datasets and more information about how the indicators are calculated and how they can be used by policymakers is available online.

ASTI

AGRICULTURAL RESEARCH INVESTMENT TRENDS

Strong, well-financed agricultural research and innovation systems are crucial for improving agricultural productivity and supporting agricultural transformation, especially in the context of ongoing global development challenges such as rapid population growth, climate change, and shifting diets. It is vital to be able to track the funding, functioning, and impact of these systems in order to improve efficiency, promote evidence-based decision-making, and demonstrate impact. Sound agricultural research investment decisions require access to up-to-date and credible qualitative and quantitative data and targeted analyses. ASTI works with national, regional, and international partners to collect time-series data on the funding, human resource capacity, and outputs of agricultural research in low- and middle-income countries. These data constitute a powerful resource for national and regional research managers, policymakers, donor organizations, and other stakeholders. The datasets and related analyses are available through a suite of tools on the ASTI website.

R&D SPENDING FALLS IN AFRICA...

Following a long period of growth, overall agricultural research expenditures (excluding the private for-profit sector) in Africa south of the Sahara (SSA) fell by 5 percent between 2014 and 2016, reflecting declining government and donor spending (Figure A). This decline in spending was widespread—about half of the African countries for which time-series data are available spent less on agricultural research in 2016 than they did in 2014. It is still unclear whether this decline signals a new trend, or was simply another instance of the funding volatility that Africa has suffered for decades.

...AND POTENTIAL PRODUCTIVITY GAINS ARE UNREALIZED

Growth in spending on agricultural research has been slower than growth in agricultural output in SSA. As a result, the region's agricultural research intensity ratio—that is, its agricultural research spending as a share of its agricultural gross domestic product (AgGDP)—dropped markedly, from 0.59 percent in 2000 to 0.39 percent in 2016. This ratio is far below the 1 percent recommended by the African Union and United Nations. For all SSA countries to meet the 1 percent target by 2030, research investment would need to increase much faster than historical growth rates. However, if the region did meet this ambitious goal, agriculture sector productivity would be 62 percent higher by 2050 than it is today, with major gains for well-being (Figure B).

FIGURE A AGRICULTURAL R&D SPENDING, TOTAL AND SHARE OF AGRICULTURAL GDP, AFRICA SOUTH OF THE SAHARA

FIGURE B PROJECTED PRODUCTIVITY GAINS IF AFRICA SOUTH OF THE SAHARA MEETS 1 PERCENT TARGET

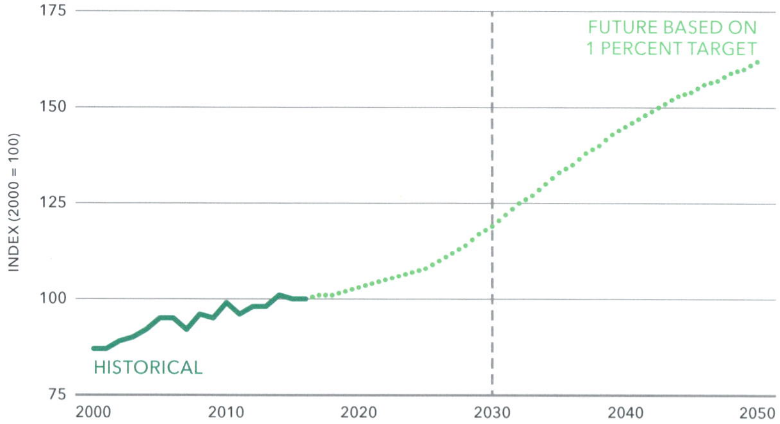

VISIT ONLINE
www.asti.cgiar.org

SPEED

TRACKING PUBLIC EXPENDITURES FOR DEVELOPMENT

The Statistics on Public Expenditures for Economic Development (SPEED) database tracks public expenditures for development, including for agriculture, to allow policymakers and analysts to examine policy priorities, track development goals, and explore the cost-effectiveness of public spending, either within a country or across countries within a region or at a similar level of development over a long timeframe. SPEED currently includes data for 166 countries in 10 public expenditure sectors from 1980 to 2017. User-friendly tools on the SPEED website enable the generation of accessible charts and geographic expenditure maps, in addition to direct data downloads.

WHAT DO RISING INCOMES MEAN FOR PUBLIC EXPENDITURE ON AGRICULTURE?

How does public spending on agriculture relate to per capita incomes? Generally speaking, the share of public spending on agriculture falls with rising income. This trend has been stronger in Latin America and the Caribbean (LAC), where income (measured as GDP per capita) has been relatively higher among developing regions. Asia also followed this trend until the 2007–08 world food price crisis. Since the crisis, the share of agricultural spending in Asia has increased modestly with income growth. The trend in Africa south of the Sahara (SSA), however, does not show a similar pattern. In SSA, incomes have risen much more slowly and the share of agricultural spending has been unsteady, with no discernible trend associated with income growth. In the early 2000s, SSA reached the same income levels that Asia had reached in the mid-1990s with comparable shares of agricultural spending. Although the share of agricultural spending has increased in Africa from its low point in the early 2000s, it remains well below the 10 percent spending target set by the Comprehensive Africa Agriculture Development Programme (CAADP).

VISIT ONLINE
SPEED webpage
https://www.ifpri.org/project/speed
SPEED database
https://dataverse.harvard.edu/dataverse/speed

FIGURE A AGRICULTURAL SHARE OF TOTAL PUBLIC SPENDING VS. LEVEL OF GDP PER CAPITA FOR ASIA, LAC, AND SSA, 1990 TO 2016

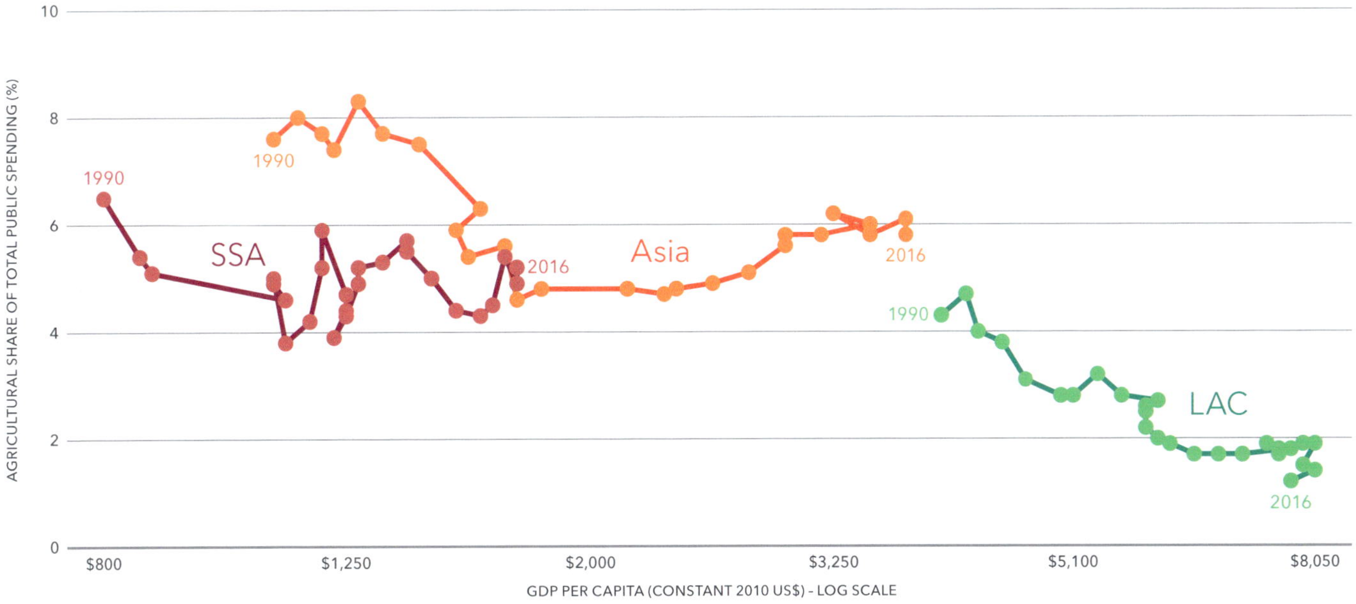

Note: Values represent simple averages across 26 developing countries in Africa south of the Sahara (SSA), 17 in Asia, and 20 in Latin America and the Caribbean (LAC).

FOOD POLICY RESEARCH CAPACITY INDICATORS

SURVEYING COUNTRY-LEVEL RESEARCH COMPETENCIES

Food policy research plays a crucial role in guiding agricultural transformation in developing countries. To achieve food security goals, countries need to strengthen their capacity to conduct food policy research. Strong local policy research institutions support evidence-based policymaking. Measuring national capacity for food policy research is important for identifying gaps and guiding the allocation of resources to fill those gaps.

"Food policy research capacity" is defined as any capacity related to socioeconomic or policy-related research in the areas of food, agriculture, nutrition, or natural resources. To measure this capacity, IFPRI has developed a set of indicators for the quantity and quality of policy research at the country level, based on data regarding PhD-level researchers and analysts (full-time equivalent) and relevant international publications. The full dataset covers 33 developing countries in Asia, Africa, and Latin America.

RESEARCH CAPACITY IS UNEVEN WITHIN DEVELOPING REGIONS

Food policy research capacity varies greatly across and within developing regions. In terms of full-time equivalent researchers with a PhD per million rural population, South Africa, Ghana, Liberia, and Togo have a relatively high ratio of food policy researchers to rural people compared with other countries in Africa south of the Sahara (see Map). Other African countries, including Mozambique, Eswatini (Swaziland), and Uganda, have relatively few researchers, suggesting lower food policy research capacity. In terms of number of publications, Ethiopia, South Africa, and Ghana have made notable progress in recent years.

DOWNLOAD DATA
https://doi.org/10.7910/DVN/XNGXTF

FIGURE A RATIO OF RESEARCHERS TO RURAL RESIDENTS, 2018

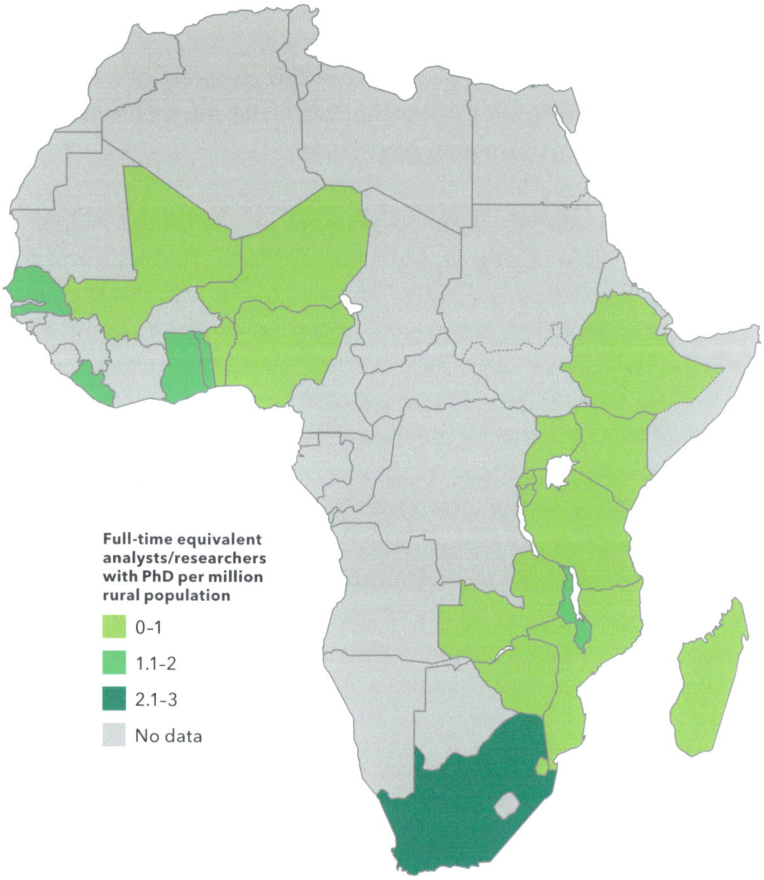

AGRICULTURAL TOTAL FACTOR PRODUCTIVITY

MEASURING EFFICIENCY IN AGRICULTURE

Increasing the efficiency of agricultural production—getting more output from the same amount of resources—is a key element of food system transformation. Total factor productivity (TFP) is an indicator of how efficiently agricultural land, labor, capital, and materials (agricultural inputs) are used to produce a country's agricultural output. It is calculated as the ratio of total agricultural output to total production inputs. Measures of land and labor productivity—partial factor productivity (PFP)—are calculated as the ratio of total output to total agricultural area (land productivity) and total output to the number of economically active persons in agriculture (labor productivity). TFP and PFP indicators contribute to the understanding of agricultural systems needed for policy and investment decisions. IFPRI calculates TFP and land and labor productivity for 132 developing countries and regions.

TFP TRENDS SHAPE PER CAPITA OUTPUT GROWTH

Growth in TFP contributes to rising incomes and greater food security. Because TFP growth is driven in the long run largely by R&D investment, the projection of TFP growth assumes that future trends in R&D investment will follow historical trends (1991–2016). Productivity growth among developing regions is diverging and will differ notably by 2050 (Figure A). TFP in East Asia (driven by China) will increase during 2017–2050 at an average rate of 1.6 percent per year, below the 2.5 percent average growth rate of 1991–2016. TFP in South and Southeast Asia is projected to grow at about 1 percent per year, a similar rate to that observed during 2001–2016. But annual TFP growth rates will drop below 1 percent for Latin America, North Africa, West and Central Asia, and Africa south of the Sahara, less than half the growth rate observed in these regions recently (2001–2016). This divergence contributes to growing differences in agricultural output.

TFP growth and increased use of inputs (land, labor, capital, and materials) together drive agricultural output. Our projections show agricultural output per capita will grow fastest in Asia, particularly in East Asia but also in Southeast and South Asia, regions where growth in R&D investment and TFP is expected to be high (Figure B). In Latin America and the Caribbean, the Middle East, Central Asia, and Africa south of the Sahara, low growth in R&D investment and consequently TFP means that growth in output per capita depends on constant increases in inputs, and output will grow more slowly than in Asia. In Africa south of the Sahara, the projected 1 percent growth in TFP per year is even slower than population growth, meaning the region will need to increase R&D investment to speed up TFP growth and increase input growth by increasing investment and the use of modern inputs.

FIGURE A TRENDS AND PROJECTIONS OF AGRICULTURAL TFP

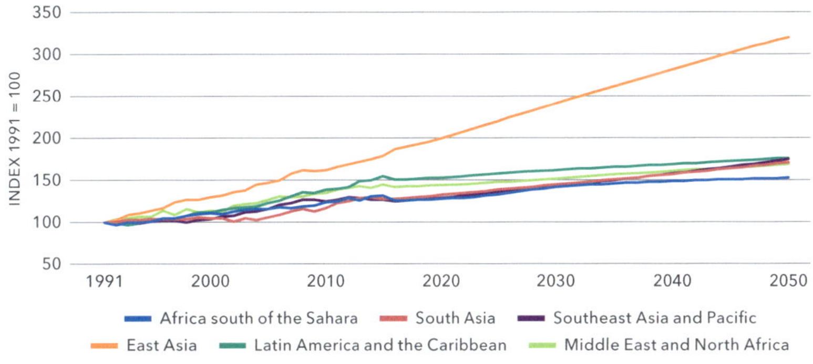

FIGURE B TRENDS AND PROJECTIONS OF AGRICULTURAL PRODUCTION PER CAPITA, WITH OUTPUT DRIVEN BY TFP AND INPUT GROWTH

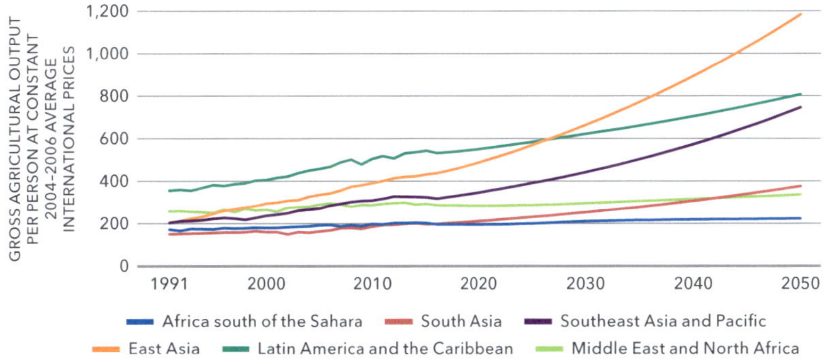

DOWNLOAD DATA
https://doi.org/10.7910/DVN/PJDGTJ

IMPACT

PROJECTIONS OF FOOD PRODUCTION, CONSUMPTION, AND HUNGER

IMPACT is an integrated system of linked economic, climate, water, and crop models that helps us explore alternative future scenarios for food and agriculture at global, regional, and national scales. Covering 158 countries, IMPACT supports in-depth analysis of a variety of critical issues of interest to policymakers, including the relationship between diets, agricultural production, and climate change. The results of alternative scenarios exploring different population, income, policy, investment, and technological pathways are available online, along with more details on the model.

HOW WILL DIETS CHANGE?

Interest in current and future diets is increasing due to the impact that diets have on human health and on the environmental footprint of agriculture. IMPACT helps us explore how the composition of diets may change under different socioeconomic and climate scenarios. The figure offers a snapshot of the average diet in 2010 and in 2050, under no-climate-change conditions (NoCC) and under climate change (RCP8.5). Looking toward 2050, continued economic development and income growth will help drive a broader transformation of the food system, including increased consumption of animal-sourced foods, fruits and vegetables, and oils and sugars. These are projected to increase their share of the average global diet, driven largely by evolving demand in developing countries. Meat consumption is expected to increase, especially in Africa and in East Asia and the Pacific. The latter will see the largest increase in oils and sugars in the diet, reflecting higher demand for processed foods, while South Asia will see the largest increase in the share of fruits and vegetables. Socioeconomic drivers appear to have a stronger effect in determining changes in future diets than does climate change—at least until 2050. Although climate change is projected to have large negative effects on the productivity of agriculture across the globe, it will have relatively little impact on average diet composition in 2050.

VISIT ONLINE

IMPACT IFPRI main page
https://www.ifpri.org/program/impact-model

IMPACT data
https://dataverse.harvard.edu/dataverse/impact

FIGURE A DIET COMPOSITION FOR AN AVERAGE CONSUMER IN 2010 AND 2050

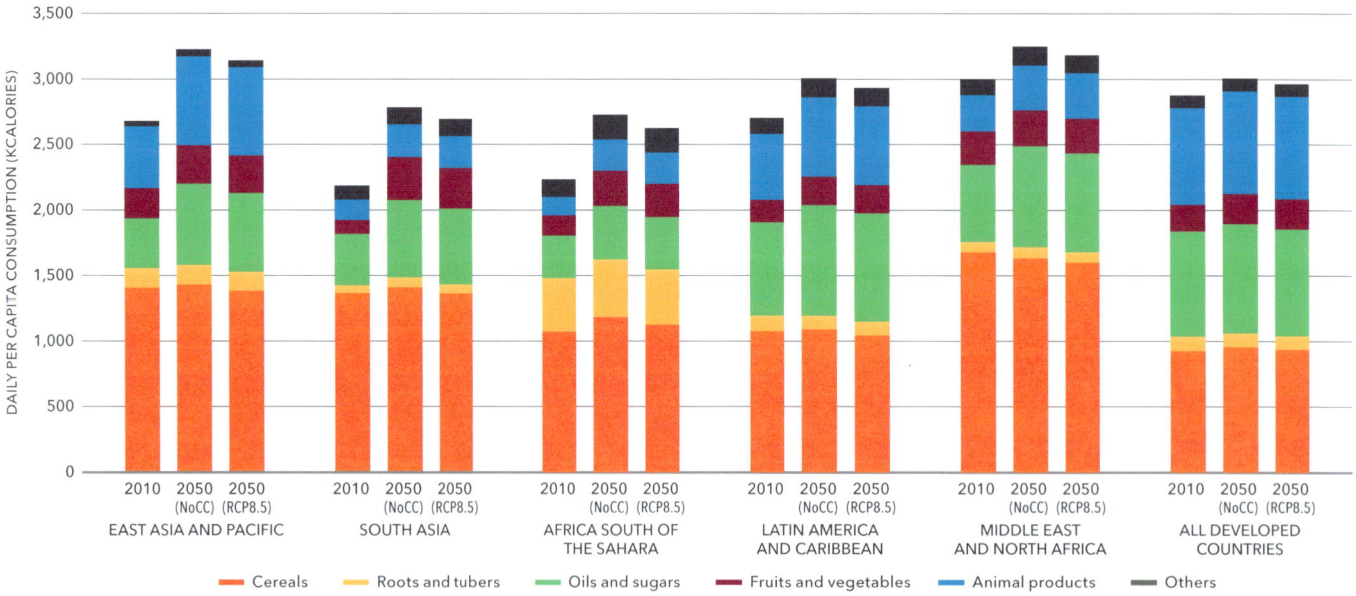

Note: Simulations are performed under assumptions of middle-of-the-road growth in population and income (Shared Socioeconomic Pathway SSP2). Data reflect the results under no climate change (NoCC), and an average of results across 5 global climate models under the Representative Concentration Pathway RCP8.5.

Notes

CHAPTER 1

1 FAO, IFAD, UNICEF, WFP, and WHO, *The State of Food Security and Nutrition in the World 2019* (Rome: FAO, 2019).

2 FAO, "Greenhouse Gas Emissions from Agriculture, Forestry and Other Land Use," 2016, http://www.fao.org/3/a-i6340e.pdf.

3 S. Lowder, J. Skoet, and T. Raney, "The Number, Size, and Distribution of Farms, Smallholder Farms, and Family Farms Worldwide," *World Development* 87 (2016): 16–29.

4 J. Njuki, C. Doss, and S. Boote, "Women's Control over Income: Implications for Women's Empowerment and the Agricultural Sector," in *2019 ReSAKSS Annual Trends and Outlook Report–Gender Equality in Rural Africa: From Commitments to Outcomes*, ed. A. Quisumbing, R. Meinzen-Dick, and J. Njuki (Washington, DC: IFPRI, 2019); C. Doss and A. Quisumbing, "Why Gender Matters for Agricultural Productivity in Africa," in *2019 ReSAKSS Annual Trends and Outlook Report*.

5 FAO, IFAD, UNICEF, WFP, and WHO, *The State of Food Security and Nutrition in the World 2017* (Rome: FAO, 2017).

6 The reach-benefit-empower framework is used by the IFPRI-led Gender, Agriculture, and Assets Project (GAAP) to conceptualize approaches to gender in agricultural development activities (http://gaap.ifpri.info/).

7 ILO (International Labour Organization), ILOSTAT database, accessed September 2019, https://data.worldbank.org/indicator/SL.AGR.EMPL.ZS.

8 G. Seymour, "Women's Empowerment in Agriculture: Implications for Technical Efficiency in Rural Bangladesh," *Agricultural Economics* 48, no. 4 (2017): 513–522; S. Sharaunga, M. Mudhara, and A. Bogale, "Effects of 'Women Empowerment' on Household Food Security in Rural KwaZulu-Natal Province," *Development Policy Review* 34, no. 2 (2016): 223–252; E. Sraboni et al., "Women's Empowerment in Agriculture: What Role for Food Security in Bangladesh?," *World Development* 61 (2014): 11–52; H. J. L. Malapit et al., "Women's Empowerment Mitigates the Negative Effects of Low Production Diversity on Maternal and Child Nutrition in Nepal," *Journal of Development Studies* 51, no. 8 (2015): 1097–1123.

9 World Inequality Lab, *World Inequality Report 2018: Executive Summary* (2017); E. Gonzalez, "Weekly Chart: Income Inequality in Latin America," Americas Society/Council of the Americas, December 8, 2017, https://www.as-coa.org/articles/weekly-chart-income-inequality-latin-america.

10 UNDP, *Human Development Report 2019: Beyond Income, Beyond Averages, Beyond Today: Inequalities in Human Development in the 21st Century* (New York: 2019).

11 L. Zhang, "An Institutional Approach to Gender Diversity and Firm Performance," *Organization Science*, forthcoming.

12 C. S. Hendrix, *Keeping Up with the Future: Upgrading Forecasts of Political Instability and Geopolitical Risk*, Policy Brief PB19-10 (Washington, DC: Peterson Institute for International Economics, 2019).

13 A. De Brauw, M. Moursi, and A. Munhaua, "Vitamin A Intakes Remain Higher among Intervention Participants 3 Years after a Biofortification Intervention in Mozambique," *British Journal of Nutrition* 122, no. 10 (2019): 1175–1181.

14 M. Maertens and J. F. M. Swinnen, "Trade, Standards, and Poverty: Evidence from Senegal," LICOS Discussion Papers, LICOS Centre for Institutions and Economic Performance (Katholieke Universiteit Leuven, Belgium, 2006).

15 S. Kurdi et al., *Responding to Conflict: Does "Cash Plus" Work for Preventing Malnutrition? New Evidence from an Impact Evaluation of Yemen's Cash for Nutrition Program*, IFPRI Policy Brief (Washington, DC: IFPRI, 2019).

16 D. O. Gilligan, J. Hoddinott, and A. S. Taffesse, "The Impact of Ethiopia's Productive Safety Net Programme and Its Linkages," *Journal of Development Studies* 45 (2009): 1684–1706; G. Tadesse, "Agriculture and Social Protection: The Experience of Ethiopia's Productive Safety Net Program," in *2017–18 ReSAKSS Annual Trends and Outlook Report–Boosting Growth to End Hunger By 2025: The Role of Social Protection*, ed. F. Wouterse and S. T. Alemayehu (Washington, DC: IFPRI, 2018).

17 A. M. Buller et al., "A Mixed-Method Review of Cash Transfers and Intimate Partner Violence in Low- and Middle-Income Countries," *World Bank Research Observer* 33, no. 2 (2018): 218–258.

18 K. Kosec and L. Wantchekon, "Can Information Improve Rural Governance and Service Delivery?" *World Development* 125 (January 2020).

19 S. Fan, "Food Policy in 2017-2018: Progress, Uncertainty, and Rising Antiglobalism," in *2018 Global Food Policy Report* (Washington, DC: IFPRI, 2018).

CHAPTER 2

1 FAO, *The State of Food and Agriculture 2017: Leveraging Food Systems for Inclusive Rural Transformation* (Rome: 2017).

2 World Bank, *Poverty and Shared Prosperity 2018: Piecing Together the Poverty Puzzle* (Washington, DC: 2018).

3 FAO, *The State of Food and Agriculture 2017.*

4 S. K. Lowder, M. V. Sanchez, and R. Bertini, "Farms, Family Farms, Farmland Distribution and Farm Labour: What Do We Know Today?" FAO Agricultural Development Economics Working Paper 19-08 (FAO, Rome, 2019).

5 Lowder, Sanchez, and Bertini, "Farms, Family Farms, Farmland Distribution and Farm Labour."

6 T. Reardon, *Growing Food for Growing Cities: Transforming Food Systems in an Urbanizing World* (Chicago: Chicago Council on Global Affairs, 2016).

7 C. Barrett et al., "Structural Transformation and Economic Development: Insights from the Agrifood Value Chain Revolution," *Journal of Economic Literature* (forthcoming); T. Reardon et al., "Overview," in *Africa Agriculture Status Report 2019: The Hidden Middle: A Quiet Revolution in the Private Sector Driving Agricultural Transformation* (Nairobi: Alliance for a Green Revolution in Africa, 2019).

8 L. Christiaensen and Y. Todo, "Poverty Reduction during the Rural-Urban Transformation—The Role of the Missing Middle," *World Development* 63 (2014): 43–58; IFAD, *Rural Development Report 2016: Fostering Inclusive Rural Transformation* (Rome: 2016); FAO, *The State of Food and Agriculture 2017.*

9 M. Dolislager et al., "Youth Agrifood System Employment in Developing Countries: A Gender-Differentiated Spatial Approach," IFAD Research Series 43 (IFAD, Rome, 2018).

10 Reardon et al., "Overview," in *Africa Agriculture Status Report 2019*; T. Reardon et al., "The Quiet Revolution in Agri-food Distribution (Wholesale, Logistics, Retail) in Sub-Saharan Africa," in *Africa Agriculture Status Report 2019*; T. Reardon et al, "The Quiet Revolution and Emerging Modern Revolution in Agri-food Processing in Sub-Saharan Africa," in *Africa Agriculture Status Report 2019.*

11 Barrett et al., "Structural Transformation and Economic Development"; Reardon et al., "Overview," in *Africa Agriculture Status Report 2019*; Reardon et al., "The Quiet Revolution in Agri-food Distribution (Wholesale, Logistics, Retail) in Sub-Saharan Africa"; Reardon et al., "Quiet Revolution and Emerging Modern Revolution."

12 G. Shively, "Infrastructure Mitigates the Sensitivity of Child Growth to Local Agriculture and Rainfall in Nepal and Uganda," *Proceedings of the National Academy of Sciences of the United States of America* 114, no. 5 (2017): 903–908.

13 FAO, *State of Food and Agriculture 2017*, 111.

14 E. Ramírez and R. Ruben, "Gender Systems and Women's Labor Force Participation in the Salmon Industry in Chiloé, Chile," *World Development* 73 (2015): 96–104.

15 H. M. Ravnborg and L. I. Gómez, "Poverty Reduction through Dispossession: The Milk Boom and the Return of the Elite in Santo Tomás, Nicaragua," *World Development* 73 (2015): 118–128.

16 G. Shively and G. Thapa, "Markets, Transportation Infrastructure and Food Prices in Nepal," *American Journal of Agricultural Economics* 99, no. 3 (2017): 660–682.

17 E. Gálvez Nogales and M. Webber, eds., *Territorial Tools for Agro-Industry Development—A Sourcebook* (Rome: FAO, 2017); E. Maruyama et al., "Frontier Analysis and Agricultural Typologies," ZEF-Discussion Papers on Development Policy No. 251 (Center for Development Research [ZEF], Bonn, Germany, 2018); and FAO, *State of Food and Agriculture 2017.*

18 IFPRI, *2019 Global Food Policy Report* (Washington, DC: 2019).

19 Reardon et al., "Overview," in *Africa Agriculture Status Report 2019*; Reardon et al., "Quiet Revolution in Agri-food Distribution"; and Reardon et al., "Quiet Revolution and Emerging Modern Revolution."

20 J. E. Snyder, "Milling Service Markets, Value Chain Coordination, and Food Attribute Values: Three Essays on the Post-farm Food System in Tanzania" (PhD diss., Michigan State University, 2018); and D. Neven et al., *Smallholder Farmer Participation in Modernization of a Food System: The Dairy Value Chain in Zambia* (Rome: FAO, 2017).

21 Reardon et al., "Quiet Revolution and Emerging Modern Revolution"; Barrett et al., "Structural Transformation and Economic Development."

22 A. Arslan, A. Mabiso, and A. Garbero, "Labor Markets during the 'Quiet Revolution': Implications for the Private Sector in the Agri-food System," in *Africa Agriculture Status Report 2019.*

23 C. Barrett, "Smallholder Market Participation: Concepts and Evidence from Eastern and Southern Africa," *Food Policy* 33, no. 4 (2008): 299–317; C. Poulton, J. Kydd, and A. Dorward, "Overcoming Market Constraints on Pro-poor Agricultural Growth in Sub-Saharan Africa," *Development Policy Review* 24, no. 3 (2006): 243–277; and T. Reardon and C. P. Timmer, "The Economics of the Food System Revolution," *Annual Review of Resource Economics* 4 (2012): 225–264.

24 K. Deininger, T. Hilhorst, and V. Songwe, "Identifying and Addressing Land Governance Constraints to Support Intensification and Land Market Operation: Evidence from 10 African Countries," *Food Policy* 48 (2014): 76–87; K. Deininger, S. Savastano, and F. Xia, "Smallholders' Land Access in Sub-Saharan Africa: A New Landscape?," *Food Policy* 67 (2017): 78–92.

25 M. F. Bellemare, "The Productivity Impacts of Formal and Informal Land Rights: Evidence from Madagascar," *Land Economics* 89, no. 2 (2013): 272–290; J. Fenske, "Land Tenure and Investment Incentives: Evidence from West Africa," *Journal of Development Economics* 95, no. 2 (2011): 137–156; S. Corsi, L. V. Marchisio, and L. Orsi, "Connecting Smallholder Farmers to Local Markets: Drivers of Collective Action, Land Tenure and Food Security in East Chad," *Land Use Policy* 68 (2017): 39–47.

26 R. Meinzen-Dick et al., "Women's Land Rights as a Pathway to Poverty Reduction: Framework and Review of Available Evidence," *World Development* 172 (2019): 72–82.

27 Lowder, Sanchez, and Bertini, "Farms, Family Farms, Farmland Distribution and Farm Labour."

28 O. Mikecz and R. Vos, "Can Smallholders Double Their Productivity and Incomes by 2030?," Agricultural Development Economics Division—ESA Working Paper No. 16-05 (FAO, Rome, 2016).

29 S. Lowder and R. Bertini, "The Transformation in the Size and Distribution of Farmland Operated by Households and Other Farms in Developing Countries: Situation, Trends, and Policies," background paper prepared for *The State of Food and Agriculture 2017* (FAO, Rome, 2017).

30 FAO, *State of Food and Agriculture 2017.*

31 FAO, *State of Food and Agriculture 2017.*

32 WLE (CGIAR Research Program on Water, Land and Ecosystems), "Enabling Sustainable, Productive Smallholder Farming Systems through Improved Land and Water Management," WLE towards Sustainable Intensification: Insights and Solutions Brief 4 (International Water Management Institute/WLE, Colombo, Sri Lanka, 2017).

33 M. Prowse, "Making Contract Farming Work with Cooperatives," ODI Opinion No. 87 (Overseas Development Institute, London, 2008).

34 D. Herbel et al., *Good Practices in Building Innovative Rural Institutions to Increase Food Security* (Rome: FAO and IFAD, 2012).

35 S. Mutonyi, "The Effect of Collective Action on Smallholder Income and Asset Holdings in Kenya," *World Development Perspectives* 14 (June 2019).

36 Y. Liu et al., "The Role of Agricultural Cooperatives in Serving as a Marketing Channel: Evidence from Low-Income Regions of Sichuan Province in China," *International Food and Agribusiness Management Review* 22, no. 2 (2019): 265–282.

37 D. Herbel et al., *Good Practices in Building Innovative Rural Institutions to Increase Food Security* (Rome: FAO and IFAD, 2012).

38 J. A. Berdegué, E. Biénabe, and L. Peppelenbos, "Conclusions: Innovative Practices in Connecting Small-Scale Producers with Dynamic Markets," in *Reconnecting Markets: Innovative Global Practices in Connecting Small-Scale Producers with Dynamic Food Markets*, ed. E. Biénabe et al. (Farnham, UK: International Institute for Environmental Development and Gower, 2011).

39 FAO, *State of Food and Agriculture 2017.*

40 GSMA, *The Mobile Economy 2019* (London: 2019).

41 GSMA, *Mobile Economy 2019.*

42 G. Conway, "Recipe for a New Revolution: Africa's Twenty-First Century Food System Transformation," *Foreign Affairs*, special issue, "African Farmers in the Digital Age" (2016): 9–16.

43 FAO and ITU (International Telecommunication Union), *E-agriculture Strategy Guide: Piloted in Asia-Pacific Countries* (Bangkok: FAO, 2016).

44 A. Adesina, "Agriculture as a Business," *Foreign Affairs*, special issue, "African Farmers in the Digital Age," 58–63.

45 FAO and ITU, *E-agriculture Strategy Guide.*

46 A. Mabiso and R. Benfica, "The Narrative on Rural Youth and Economic Opportunities in Africa: Facts, Myths and Gaps," 2019 Rural Development Report Background Paper No. 19 (IFAD, Rome, 2019).

47 H. Alderman and R. Yemtsov, "How Can Safety Nets Contribute to Economic Growth?," *World Bank Economic Review* 28, no. 1 (2014): 1-20; FAO, *The State of Food and Agriculture 2015: Social Protection and Agriculture: Breaking the Cycle of Rural Poverty* (Rome: 2015); N. Tirivayi, M. Knowles, and B. Davis, "The Interaction between Social Protection and Agriculture: A Review of Evidence," *Global Food Security* 10 (September 2016): 52-62.

48 S. Daidone et al., "Linking Agriculture and Social Protection for Food Security: The Case of Lesotho," *Global Food Security* 12 (March 2017): 146-154.

49 S. Singh and M. Fernandes, "Home-Grown School Feeding: Promoting Local Production Systems Diversification through Nutrition Sensitive Agriculture," *Food Security* 10, no. 1 (2018): 111-119.

50 D. O. Gilligan, J. Hoddinott, and A. S. Taffesse, "The Impact of Ethiopia's Productive Safety Net Programme and Its Linkages," *Journal of Development Studies* 45 (2009): 1684-1706; G. Tadesse, "Agriculture and Social Protection: The Experience of Ethiopia's Productive Safety Net Program," in *2017-18 ReSAKSS Annual Trends and Outlook Report-Boosting Growth to End Hunger by 2025: The Role of Social Protection*, ed. F. Wouterse and S. T. Alemayehu (Washington, DC: IFPRI, 2018).

51 M. Filipski et al., "General Equilibrium Impact Assessment of the Productive Safety Net Program in Ethiopia," mimeo, Grantee Final Report to 3ie (IFPRI, 2016); M. Arega and G. Shively, "Food Aid, Cash Transfers and Producer Prices in Ethiopia," *African Journal of Agricultural and Resource Economics* 14, no. 3 (2019): 153-168.

CHAPTER 3

1 This is based on the UN definitions of youth and the working-age population: people aged 15-24 and 15-64 years old, respectively. African countries often adopt a broader definition of youth: people aged 15-34.

2 D. Resnick and J. Thurlow, eds., *African Youth and the Persistence of Marginalization: Employment, Politics, and Prospects for Change* (New York: Routledge, 2015).

3 Some countries in regions that developed earlier addressed the needs of their youth populations by investing in educational programs and implementing import substitution or export promotion strategies to encourage local manufacturing and services. The most successful strategies in this era required substantial government resources and coordination across sectors, which may preclude similar efforts in Africa.

4 J. Thurlow, B. Davies, and P. Dorosh, "Demographic Change, Agriculture and Rural Poverty," in *Sustainable Food and Agriculture: An Integrated Approach*, ed. C. Campanhola and S. Pandey (London: Elsevier Academic Press, 2018).

5 P. Dorosh and J. Thurlow, "Can Cities or Towns Drive African Development? Economywide Analysis for Ethiopia and Uganda," *World Development* 63, no. 10 (2014): 113-123.

6 D. Filmer and L. Fox, *Youth Employment in Sub-Saharan Africa* (Washington, DC: World Bank, 2014); IFAD, *2019 Rural Development Report: Creating Opportunities for Rural Youth* (Rome: 2019).

7 J. Chamberlin, D. D. Headey, and T. S. Jayne, "Land Pressures, the Evolution of Farming Systems, and Development Strategies in Africa: A Synthesis," *Food Policy* 48 (October, 2014): 1-17.

8 S. Haggblade, D. C. Mead, and R. L Meyer, "An Overview of Programs for Promoting the Rural Nonfarm Economy," in *Transforming the Rural Nonfarm Economy: Opportunities and Threats in the Developing World*, ed. S. Haggblade, P. B. R. Hazell, and T. Reardon (Baltimore: Johns Hopkins University Press, 2007).

9 E. B. McCullough, "Labor Productivity and Employment Gaps in Sub-Saharan Africa," *Food Policy* 67 (2017): 133-152.

10 V. Mueller and J. Thurlow, eds., *Youth and Jobs in Rural Africa: Beyond Stylized Facts* (Oxford: Oxford University Press and IFPRI, 2019).

11 E. Schmidt and F. B. Woldeyes, "Rural Youth and Employment in Ethiopia," in *Youth and Jobs in Rural Africa*.

12 T. Benson, A. Erman, and B. Baulch, "Change and Rigidity in Youth Employment Patterns in Malawi," in *Youth and Jobs in Rural Africa*.

13 X. Diao et al., "Cities and Rural Transformation: A Spatial Analysis of Rural Youth Livelihoods in Ghana," in *Youth and Jobs in Rural Africa*.

14 X. Diao, E. Magalhaes, and M. McMillan, "Rural Nonfarm Enterprises in Tanzania's Economic Transformation: The Role of the Youth," in *Youth and Jobs in Rural Africa*.

15 E. Estruch et al., "Youth Mobility and Its Role in Structural Transformation in Senegal," in *Youth and Jobs in Rural Africa*.

16 Ø. Kravdal, "Education and Fertility in Sub-Saharan Africa: Individual and Community Effects," *Demography* 39 (2002): 233-250.

17 V. Mueller and H. L. Lee, "Can Migration Be a Conduit for Transformative Youth Employment?" in *Youth and Jobs in Rural Africa*.

18 D. Schwebel et al., "Policies for Youth Employment in Sub-Saharan Africa," in *Youth and Jobs in Rural Africa*.

19 D. Resnick, "Troublemakers, Bystanders, and Pathbreakers: The Political Participation of African Youth," in *Youth and Jobs in Rural Africa*.

20 Schwebel et al., "Policies for Youth Employment in Sub-Saharan Africa," in *Youth and Jobs in Rural Africa*.

CHAPTER 4

1 W. M. H. Jaim and M. Hossain, "Women's Participation in Agriculture in Bangladesh 1988-2008: Changes and Determinants," paper presented in preconference event on Dynamics of Rural Livelihoods and Poverty in South Asia, 7th Asian Society of Agricultural Economists International Conference (Hanoi, Viet Nam, 2011); S. K. Mutavi et al., "Innovativeness and Adaptations: The Way Forward for Small Scale Peri-urban Dairy Farmers in Semi-arid Regions of South Eastern Kenya," *International Journal of Scientific Research and Innovative Technology* 3, no. 5 (2016): 1-14; H. Takeshima et al., "Agriculture-Nutrition Linkages, Cooking-Time, Intrahousehold Equality among Women and Children: Evidence from Tajikistan," IFPRI Discussion Paper 1882 (IFPRI, Washington, DC, 2019).

2 C. Doss and A. Quisumbing, "Why Gender Matters for Agricultural Productivity in Africa," in *2019 ReSAKSS Annual Trends and Outlook Report-Gender Equality in Rural Africa: From Commitments to Outcomes*, ed. A. Quisumbing, R. Meinzen-Dick, and J. Njuki (Washington, DC: IFPRI, 2019); J. Njuki, C. R. Doss, and S. Boote, "Women's Control over Income: Implications for Women's Empowerment and the Agricultural Sector," in *2019 ReSAKSS Annual Trends and Outlook Report*.

3 C. Doss et al., "Gender, Rural Youth and Structural Transformation: Evidence to Inform Innovative Youth Programming," 2019 Rural Development Report Background Paper No. 4 (IFAD, Rome, 2019); D. Rubin, B. Boonabaana, and C. Manfre, "Building an Inclusive Agriculture: Strengthening Gender Equality in Agricultural Value Chains," in *2019 ReSAKSS Annual Trends and Outlook Report*.

4 N. Kabeer, "Resources, Agency, Achievements: Reflections on the Measurement of Women's Empowerment," *Development and Change* 30, no. 3 (1999): 435–464.

5 N. Johnson et al., "How Do Agricultural Development Projects Empower Women? Linking Strategies with Expected Outcomes," *Journal of Gender, Agriculture and Food Security* 3, no. 2 (2018): 1–19.

6 S. Theis and R. Meinzen-Dick, "Reach, Benefit, or Empower: Clarifying Gender Strategies of Development Projects," *IFPRI Blog*, December 16, 2016, https://www.ifpri.org/blog/reach-benefit-or-empower-clarifying-gender-strategies-development-projects.

7 IFPRI (website), "About WEAI," accessed December 12, 2019, http://weai.ifpri.info/about-weai/.

8 A. Quisumbing et al., "Gender, Assets, and Market-Oriented Agriculture: Learning from High-Value Crop and Livestock Projects in Africa and Asia," *Agriculture and Human Values* 32, no. 4 (2015): 705–725.

9 M. Said-Allsopp and A. Tallontire, "Pathways to Empowerment?: Dynamics of Women's Participation in Global Value Chains," *Journal of Cleaner Production* 107 (November 2015): 114-121; and S.Hanson, "Changing Places through Women's Entrepreneurship," *Economic Geography* 85, no. 3 (2015): 245–267.

10 A. Ahmed et al., *Tracking Empowerment along the Value Chain: Testing a Modified WEAI in the Feed the Future Zone of Influence in Bangladesh* (Dhaka, Bangladesh, and Washington, DC: IFPRI, 2018); H. Malapit et al., "Empowerment in Agricultural Value Chains: Mixed Methods Evidence from the Philippines," IFPRI Discussion Paper 1881 (IFPRI, Washington, DC, 2019).

11 G. Seymour, "Women's Empowerment in Agriculture: Implications for Technical Efficiency in Rural Bangladesh," *Agricultural Economics* 48, no. 4 (2017): 513–522; S. Sharaunga, M. Mudhara, and A. Bogale, "Effects of 'Women Empowerment' on Household Food Security in Rural KwaZulu-Natal Province," *Development Policy Review* 34, no. 2 (2016): 223–252; E. Sraboni et al., "Women's Empowerment in Agriculture: What Role for Food Security in Bangladesh?" *World Development* 61 (2014): 11–52; and H. Malapit et al., "Women's Empowerment Mitigates the Negative Effects of Low Production Diversity on Maternal and Child Nutrition in Nepal," *Journal of Development Studies* 51, no. 8 (2015): 1097–1123.

12 J. Twyman et al., "Gender Equity Considerations for Healthier Diets Research: A Systematic Review" (unpublished manuscript for A4NH, Washington, DC, 2019).

13 V. Slavchevska, S. Kaaria, and S.-L. Taivalmaa, *Feminization of Agriculture in the Context of Rural Transformations: What Is the Evidence?* (Washington, DC: World Bank, 2016); A. Maharjan, S. Bauer, and B. Knerr, "Do Rural Women Who Stay Behind Benefit from Male Out-Migration? A Case Study in the Hills of Nepal," *Gender, Technology and Development* 16, no. 1 (2012): 95–123.

14 K. Ambler, K. Jones, and M. O'Sullivan, "What Is the Role of Men in Connecting Women to Cash Crop Markets? Evidence from Uganda," IFPRI Discussion Paper 1762 (IFPRI, Washington, DC, 2018).

15 GROOTS Kenya, "Mainstreaming Climate Change and Disaster Risk Reduction into the Laikipia County CIDP 2018-2022," Multi-stakeholder Action Planning Report (Nairobi, 2019).

16 N. Kumar et al., "Social Networks, Mobility, and Political Participation: The Potential for Women's Self-Help Groups to Improve Access and Use of Public Entitlement Schemes in India," *World Development* 114 (2019): 28–41.

17 R. Meinzen-Dick et al., "Women's Land Rights as a Pathway to Poverty Reduction: Framework and Review of Available Evidence," *Agricultural Systems* 172 (2019): 72–82.

18 J. Njuki et al., "Beyond Access: Gender-Transformative Financial Inclusion in Agriculture and Entrepreneurship," in *2019 ReSAKSS Annual Trends and Outlook Report*.

19 S. Theis et al., "What Happens after Technology Adoption? Gendered Aspects of Small-Scale Irrigation Technologies in Ethiopia, Ghana, and Tanzania," *Agriculture and Human Values* 35, no. 3 (2018): 671–684.

20 A. Tall et al., *Scaling Up Climate Services for Farmers–Mission Possible: Learning from Good Practice in Africa and South Asia*, CCAFS Report No. 13 (Wageningen, Netherlands: CGIAR Research Program on Climate Change, Agriculture and Food Security, 2014).

21 A. Quisumbing et al., "Closing the Gender Asset Gap: Learning from Value Chain Development in Africa and Asia," IFPRI Discussion Paper 1321 (IFPRI, Washington, DC, 2014).

22 D. Rubin, B. Boonabaana, and C. Manfre, "Building an Inclusive Agriculture: Strengthening Gender Equality in Agricultural Value Chains," in *2019 ReSAKSS Annual Trends and Outlook Report*.

23 IFPRI (website), "WEAI4VC," accessed November 6, 2019, http://weai.ifpri.info/versions/weai4vc/.

24 ISEAL Alliance (website), "Gender and Sustainability Standards," accessed November 7, 2019, https://www.isealalliance.org/about-iseal/our-work/gender-working-group-sustainability-standards.

25 L. Forsythe, H. Posthumus, and A. Martin, "A Crop of One's Own? Women's Experiences of Cassava Commercialization in Nigeria and Malawi," *Journal of Gender, Agriculture, and Food Security* 1, no. 2 (2016): 110–128.

26 S. Lyon, T. Mutersbaugh, and H. Worthen, "The Triple Burden: The Impact of Time Poverty on Women's Participation in Coffee Producer Organizational Governance in Mexico," *Agriculture and Human Values* 34, no. 2 (2017): 317–331.

27 R. S. Meinzen-Dick et al., "Women's Empowerment in Agriculture: Lessons from Qualitative Research," IFPRI Discussion Paper 1797 (IFPRI, Washington, DC, 2019).

28 Njuki et al., "Beyond Access."

CHAPTER 5

1 C. Holleman et al., *Sowing the Seeds of Peace for Food Security: Disentangling the Nexus between Conflict, Food Security and Peace*, FAO Agricultural Development Economics Technical Study No. 2 (Rome: FAO, 2017); FAO, IFAD, UNICEF, WFP, and WHO, *The State of Food Security and Nutrition in the World 2017* (Rome: FAO, 2017).

2 N. Crawford et al., *Protracted Displacement: Uncertain Paths to Self-Reliance in Exile* (London: ODI Humanitarian Policy Group, 2015).

3 UNHCR, *Global Trends: Forced Displacement in 2018* (Geneva: 2019).

4 Holleman et al., *Sowing the Seeds of Peace*.

5 UNHCR, "Operational Update: Yemen," October 11, 2019; FAO website, "FAO in Emergencies: Crisis in Yemen," accessed December 5, 2019, http://www.fao.org/emergencies/crisis/yemen/en/.

6 FAO, IFAD, UNICEF, WFP, and WHO, *State of Food Security and Nutrition in the World 2017*.

7 FAO, *The State of Food and Agriculture 2018: Migration, Agriculture and Rural Development* (Rome: 2018).

8 FAO, *FAO Migration Framework: Migration as a Choice and an Opportunity for Rural Development* (Rome: 2019).

9 World Bank, *Forcibly Displaced: Toward a Development Approach Supporting Refugees, the Internally Displaced, and Their Hosts* (Washington, DC: 2017); Holleman et al., *Sowing the Seeds of Peace*.

10 FAO, *Counting the Cost: Agriculture in Syria after Six Years of Crisis* (Rome: 2017); UNHCR, "Syria Factsheet, January 2019.

11 Holleman et al., *Sowing the Seeds of Peace*.

12 O. Ecker, "Resilience for Food Security in the Face of Civil Conflict in Yemen," in *Resilience for Food and Nutrition Security*, ed. S. Fan, R. Pandya-Lorch, and S. Yosef (Washington, DC: IFPRI, 2014).

13 C. Breisinger, O. Ecker, and J. F. Trinh Tan, "Conflict and Food Insecurity: How Do We Break the Links?," in *Global Food Policy Report 2014-15* (Washington, DC: IFPRI, 2015); Holleman et al., *Sowing the Seeds of Peace*.

14 FAO IFAD, UNICEF, WFP, and WHO, *State of Food Security and Nutrition in the World 2017*.

15 Holleman et al., *Sowing the Seeds of Peace*; FAO and IFPRI, *Conflict, Migration and Food Security: The Role of Agriculture and Rural Development*, Joint Policy Brief (Rome and Washington, DC: 2017).

16 For example, the UN and the World Bank set up the New Way of Working (NWoW) to deliver the nexus approach (see Agenda for Humanity: New Way of Working, https://www.agendaforhumanity.org/initiatives/5358). The OECD has made the nexus a priority, and members of the Development Assistance Committee have started adjusting funding mechanisms for crisis-response programs (https://legalinstruments.oecd.org/public/doc/643/643.en.pdf). All UN agencies and many donors and NGOs support the approach.

17 FAO, IFAD, UNICEF, WFP, and WHO, *State of Food Security and Nutrition in the World 2017*.

18 UNHCR, "Refugee Statistics," accessed December 5, 2019, https://www.unrefugees.org/refugee-facts/statistics/.

19 FAO reports that both returning and remaining refugees in Uganda developed strong economic ties with local communities, generating new employment for local Ugandans as well; FAO, *State of Food and Agriculture 2018*; FAO and Office of the Prime Minister of Uganda, *Food Security, Resilience and Well-Being Analysis of Refugees and Host Communities in Northern Uganda* (Rome: 2018). Betts et al. note that this spillover effect was only partly due to the humanitarian assistance provided. Most refugees also relied on their social networks to support the development of their new livelihoods; A. Betts et al., *Refugee Economies: Rethinking Popular Assumptions* (Oxford: University of Oxford Refugees Studies Centre, 2014).

20 The policy is reflective of Ethiopia's commitments under UNHCR's CRRF and the Intergovernmental Authority on Development's Declaration on Durable Solutions, which seeks to move toward durable solutions for Somali refugees.

21 UNHCR, *Global Trends*.

22 A. Segovia, "The Relationships between Food Security and Violent Conflicts: The Case of Colombia," ESA Agricultural Development Economics Working Paper 17-06 (FAO, Rome, 2017); Holleman et al., *Sowing the Seeds of Peace*.

23 Cash+ approaches integrate two complementary modes of intervention, cash transfers and productive assistance, in order to restore resilient livelihoods over the longer term. See FAO, *The Cash+ Approach in the Sahel as a Tool to Reinforce Resilience: Mali/Mauritania* (Dakar, 2017).

24 S. Kurdi, Y. Ghorpade, and H. Ibrahim, "The Cash for Nutrition Intervention in Yemen: Impact Evaluation Study," IFPRI Middle East and Northern Africa Regional Program Working Paper (IFPRI, Cairo, 2019).

25 FAO, *State of Food and Agriculture 2018*.

CHAPTER 6

1 See A4NH Flagship 1 (http://a4nh.cgiar.org/our-research/flagship-1/) and A4NH website (http://a4nh.cgiar.org/).

2 This classification is similar to that offered by the World Bank, *The Safe Food Imperative: Accelerating Progress in Low- and Middle-Income Countries* (Washington, DC: 2018). It is also consistent with others: R. Nugent et al., "Indicators for Nutrition-Friendly and Sustainable Food Systems," in *Global Nutrition Report 2015: Actions and Accountability to Advance Nutrition and Sustainable Development* (Washington, DC: IFPRI, 2015) and HLPE (High-Level Panel of Experts), *Nutrition and Food Systems: A Report by the High-Level Panel of Experts on Food Security and Nutrition* (Rome: 2017).

3 HLPE, *Nutrition and Food Systems*.

4 Global Panel on Agriculture and Food Systems for Nutrition (Glopan), *Food Systems and Diets: Facing the Challenges of the 21st Century* (London: 2016); HLPE, *Nutrition and Food Systems*.

5 FAO (website), "Food-Based Dietary Guidelines," http://www.fao.org/nutrition/education/food-dietary-guidelines/home/en/.

6 D. Tschirley et al., "The Rise of a Middle Class in East and Southern Africa: Implications for Food System Transformation," *Journal of International Development* 27, no. 5 (2015): 628-646.

7 U. Shrivastava et al., "Obesity, Diabetes, and Cardiovascular Diseases in India: Public Health Challenges," *Current Diabetes Reviews* 13 (2017): 65.

8 B. Popkin, C. Coravalan, and I. M. Grummer-Straman, "Dynamics of the Double Burden of Malnutrition and the Changing Nutrition Reality," *Lancet* 395 (2020): 65-74.

9 B. A. Swinburn et al., "The Global Syndemic of Obesity, Undernutrition, and Climate Change: The *Lancet* Commission Report," *Lancet* 393, no. 10173 (2019): 791-846.

10 B. Popkin, "Relationship between Shifts in Food System Dynamics and Acceleration of the Global Nutrition Transition," *Nutrition Reviews* 75, no. 2 (2017): 73-82.

11 Popkin, "Relationship between Shifts in Food System Dynamics."

12 E. Nakasone, M. Torero, and B. Minten, "The Power of Information: The ICT Revolution in Agricultural Development," *Annual Review of Resource Economics* 6 (2014): 533-550; J. Aker, I. Ghosh, and J. Burrell, "The Promise (and Pitfalls) of ICT for Agricultural Initiatives," *Agricultural Economics* 47, no. S1 (2016): 35-48.

13 T. Bernard et al., "Expected Product Market Reforms and Technology Adoption by Senegalese Onion Producers," *American Journal of Agricultural Economics* 99, no. 4 (2017): 1096-1115.

14 C. Saenger, M. Torero, and M. Qaim, "Impact of Third-Party Contract Enforcement in Agricultural Markets–A Field Experiment in Vietnam," *American Journal of Agricultural Economics* 96, no. 4 (2014): 1220-1238.

15 D. Donaldson, "Railroads of the Raj: Estimating the Impact of Transportation Infrastructure," *American Economic Review* 108, no. 4-5 (2018): 899-934; S. Asher and P. Novosad, "Rural Roads and Local Economic Development," Policy Research Working Paper No. 8466 (World Bank, Washington, DC, 2019).

16 B. Minten et al., "The New and Changing Roles of Cold Storages in the Potato Supply Chain in Bihar," *Economics & Political Weekly* 49, no. 52 (2014): 1-37.

17 H. E. Bouis, P. Eozenou, and A. Rahman, "Food Prices, Household Income, and Resource Allocation: Socioeconomic Perspectives on Their Effects on Dietary Quality and Nutritional Status," *Food and Nutrition Bulletin* 32, no. 1 Suppl (2011): S14-S23.

18 D. Headey and H. Alderman, "The Relative Caloric Prices of Healthy and Unhealthy Foods Differ Systematically across Income Levels and Continents," *Journal of Nutrition* 149, no. 11 (2019): 2020–2033.

19 P. Pingali et al., *Transforming Food Systems for a Rising India* (Cham, Switzerland: Palgrave Macmillan, 2019).

20 Bouis et al., "Food Prices, Household Income, and Resource Allocation."

21 World Bank, *Future of Food: Shaping the Food System to Deliver Jobs* (Washington, DC: 2017).

22 D. Filmer and L. Fox, *Youth Employment in Sub-Saharan Africa* (Washington, DC: World Bank, 2014).

23 T. Roepstorff, S. Wiggins, and A. Hawkins, "The Profile of Agribusiness in Africa," in *Agribusiness for Africa's Prosperity* (Vienna: United Nations Industrial Development Organization, 2011).

24 L. Pritchett, "Is India a Flailing State? Detours on the Four Lane Highway to Modernization," Scholarly Articles 4449106, Harvard Kennedy School of Government, 2009.

25 EUFIC (European Food Information Council), Global Update on Nutrition Labelling (2018 edition), https://www.eufic.org/images/uploads/healthy-living/Executive-Summary-GUNL-2018-V2.pdf.

26 B. Popkin, "Confronting the New Face of Malnutrition: Regulatory and Fiscal Approaches to Improving Diets," 28th Annual Martin J. Forman Memorial Lecture, October 29, 2018, IFPRI, Washington, DC.

27 See, for example, the Institute for Health Metrics and Evaluation (www.healthdata.org/data-visualization/lbd-u5m).

28 See, for example, the Women's Empowerment in Agriculture Index (https://weai.ifpri.info/).

29 See, for example, H. Kharas, J. W. McArthur, and I. Ohno, eds., *Leave No One Behind: Time for Specifics on the Sustainable Development Goals* (Washington, DC: Brookings Institute, 2019).

30 World Bank, *Future of Food: Shaping a Climate-Smart Global Food System* (Washington, DC: 2015); World Bank, *Future of Food: Shaping the Global Food System to Deliver Improved Nutrition and Health* (Washington, DC: 2016); World Bank, *Future of Food: Shaping the Food System to Deliver Jobs* (Washington, DC: 2017); World Bank, *Future of Food: Harnessing Digital Technologies to Improve Food System Outcomes* (Washington, DC: 2019).

REGIONAL DEVELOPMENTS

AFRICA

1 This is based on data for Africa as a whole. ReSAKSS (Regional Strategic Analysis and Knowledge Support System) database, accessed December 10, 2019, https://www.resakss.org/node/11.

2 A. Bouët and S. Odjo, eds., *Africa Agriculture Trade Monitor 2019* (Washington, DC: IFPRI, 2019).

3 African Union Commission, *Malabo Declaration on Accelerated Agricultural Growth and Transformation for Shared Prosperity and Improved Livelihoods* (Addis Ababa: 2014).

4 Malawi, Ministry of Agriculture, Irrigation, and Water Development, *National Agricultural Investment Plan; Prioritized and Coordinated Agricultural Transformation Plan for Malawi; FY 2017/18–2022/23* (Lilongwe: 2018).

5 World Bank, World Development Indicators database, 2019.

6 W. A. Masters et al., "Measuring the Affordability of Nutritious Diets in Africa: Price Indexes for Diet Diversity and the Cost of Nutrient Adequacy," *American Journal of Agricultural Economics* 100, no. 5 (2018): 1285–1301.

7 K. Bird, A. McKay, and I. Shinyekwa, "Isolation and Poverty: The Relationship between Spatially Differentiated Access to Goods and Services and Poverty," ODI Working Paper 322/CPRC Working Paper 162 (Overseas Development Institute, London, 2010).

8 G. Shively, "Infrastructure Mitigates the Sensitivity of Child Growth to Local Agriculture and Rainfall in Nepal and Uganda," *Proceedings of the National Academy of Sciences of the United States of America (PNAS)* 114, no. 5 (2017): 903–908.

9 F. J. Proctor and J. A. Berdegué, "Food Systems at the Rural-Urban Interface," Working Paper Series 194 (Rimisp, Santiago, Chile, 2016).

10 Malabo Montpellier Panel, *Byte by Byte: Policy Innovation for Transforming Africa's Food System with Digital Technologies* (Dakar: 2019).

11 World Bank, World Development Indicators database, 2019.

12 Malabo Montpellier Panel, *Mechanized: Transforming Africa's Agriculture Value Chains* (Dakar: 2018).

13 World Bank Group, *Africa's Pulse*, vol. 16, *An Analysis of Issues Shaping Africa's Economic Future* (Washington, DC: 2017).

14 African Union, *The African Women's Decade–Theme: Grassroots Approach to Gender Equality and Women's Empowerment (GEWE)* (Addis Ababa: 2011).

15 SADC Gender Protocol Alliance, "The SADC Protocol and Post 2015 Agenda," accessed December 13, 2019, https://genderlinks.org.za/what-we-do/sadc-gender-protocol/the-sadc-gender-protocol/.

16 Burkina Faso, "Campagne de Confinement de la Houe Manuelle au Musée: Un Forum pour Réfléchir sur l'Autonomisation Économique de la Femme," October 14, 2019.

17 R. Meinzen-Dick, E. C. Myers, and A. Quisumbing, "How Empowered Are Women in African Agriculture?" in *2019 ReSAKSS Annual Trends and Outlook Report–Gender Equality in Rural Africa: From Commitments to Outcomes*, ed. A. Quisumbing, R. Meinzen-Dick, and J. Njuki (Washington, DC: IFPRI, 2019).

18 C. Doss and A. Quisumbing, "Why Gender Matters for Agricultural Productivity in Africa," in *2019 ReSAKSS Annual Trends and Outlook Report*.

19 D. Rubin, B. Boonabaana, and C. Manfre, "Building an Inclusive Agriculture: Strengthening Gender Equality in Agricultural Value Chains," in *2019 ReSAKSS Annual Trends and Outlook Report*.

20 Rwanda Ministry of Agriculture and Animal Resources, *Strategic Plan for Agriculture Transformation, Phase 4 (PSTA 4)-2018–2024* (Kigali: 2018).

21 G. Tadesse, "Agriculture and Social Protection: The Experience of Ethiopia's Productive Safety Net Program," in *2017–2018 ReSAKSS Annual Trends and Outlook Report-Boosting Growth to End Hunger by 2025: The Role of Social Protection*, ed. F. Wouterse and A. S. Taffesse (Washington, DC: IFPRI, 2018).

22 W. Marivoet, J. Ulimwengu, and F. Sedano, "Spatial Typology for Targeted Food and Nutrition Security Interventions," *World Development* 120 (August 2019): 62–75.

MIDDLE EAST AND NORTH AFRICA

1 IMF (International Monetary Fund), "Regional Economic Outlook: Middle East and Central Asia Update," April 2019.

2 World Bank, *Reaching New Heights: Promoting Fair Competition in the Middle East and North Africa*, MENA Economic Update (Washington, DC: 2019).

3 Economist Intelligence Unit (EIU) country reports for Bahrain, Saudi Arabia, and Iraq, all from September 2019.

4 IMF, "Regional Economic Outlook: Middle East and Central Asia Update."

5 EBRD (European Bank for Reconstruction and Development), EIB (European Investment Bank), and World Bank, *What's Holding Back the Private Sector in MENA? Lessons from the Enterprise Survey* (London, Luxembourg, and Washington, DC: 2016).

6 EIU country reports for Saudi Arabia, Jordan, Kuwait, Egypt, and Algeria, all from September 2019.

7 EIU country reports for Saudi Arabia, Bahrain, Kuwait, Qatar, and Oman, all from September 2019; IMF, *Regional Economic Outlook: Middle East and Central Asia* (Washington, DC: 2018).

8 K. Elborgh-Woytek et al., "Women, Work, and the Economy: Macroeconomic Gains from Gender Equity," IMF Staff Discussion Note 13/10 (Washington, DC, 2013).

9 C. Breisinger et al., "Egypt's Takaful and Karama Cash Transfer Program: Evaluation of Program Impacts and Recommendations," IFPRI Policy Brief (Washington, DC: IFPRI, 2018).

10 M. Keulertz, "Water and Food Security Strategies in the MENA Region," MENARA Future Notes No. 18 (Barcelona: Middle East and North Africa Regional Architecture, 2019).

11 Jordan is an exception as most agricultural labor is "foreign labor" from countries such as Egypt and not counted in the underlying statistic.

12 McKinsey, "Lebanon Economic Vision," accessed December 18, 2019, https://www.economy.gov.lb/media/11893/20181022-1228full-report-en.pdf?fbclid=IwAR1D3dyuF33kZKS07CWomIljvLPFaOpA4Rb213Ve3PDbhcHF6ijsdcqRIQA.

13 ESCWA (United Nations Economic and Social Commission for Western Asia), "Rethinking Fiscal Policy for the Arab Region," E/ESCWA/EDID/2017.4 (Beirut, Lebanon, 2017).

14 World Economic Forum, *The Future of Jobs and Skills in the Middle East and North Africa: Preparing the Region for the Fourth Industrial Revolution* (Geneva: 2017).

15 D. Moshashai, "The Middle East Needs Radical Agricultural Transformation," Castlereagh Associates, July 26, 2019; FAO and UNIDO (United Nations Industrial Development Organization), *Agroindustries for Development* (Wallingford, UK: 2009).

16 A. M. Althibah et al., "Yemen: Economy-Wide Impact of Conflict and Alternative Scenarios for Recovery," Regional Program Policy Note 2 (Washington, DC, and Cairo, Egypt: IFPRI, 2019).

17 World Bank, *World Development Report 2008: Agriculture for Development* (Washington, DC: 2008).

CENTRAL ASIA

1 Uzbekistan, Ministry of Employment and Labor Relations, February 4, 2020, https://mehnat.uz/uz/news/2019-yilda-uzbekistonning-ishsizlik-darazhasi-qancha.

2 Authors' calculations using data on unemployment from Kyrgyzstan, National Statistical Committee, "Employment Statistics," accessed January 2, 2020, http://stat.kg/en/statistics/zanyatost/.

3 Data from the national statistical agencies of Central Asian countries suggest that the food processing sector's contribution to GDP is significantly lower than that of agriculture.

4 H. Takeshima et al., "Agriculture-Nutrition Linkages in Tajikistan: Evidence from Household Survey Data," IFPRI Discussion Paper 1770 (IFPRI, Washington, DC, 2018); H. Takeshima et al., "Agriculture-Nutrition Linkages, Cooking Time, Intrahousehold Equality among Women and Children: Evidence from Tajikistan,"IFPRI Discussion Paper 1882 (IFPRI, Washington, DC, 2019); K. Kosec et al., "Aspirations and Women's Empowerment: Evidence from Kyrgyzstan," IFPRI Discussion Paper 1786 (IFPRI, Washington, DC, 2018).

5 R. Chami, C. Fullenkamp, and S. Jahjah, *Are Immigrant Remittance Flows a Source of Capital for Economic Development?*, IMF Staff Papers, vol. 52, no. 1 (Washington, DC: 2005); D. Ratha, S. Mohapatra, and E. Scheja, *Impact of Migration on Economic and Social Development*, World Bank Policy Research Paper 5558 (Washington, DC: 2011).

6 IFPRI, *Feed the Future Tajikistan 2015: Zone of Influence Interim Assessment Report* (Washington, DC: 2016).

7 FAO, IOM (International Office on Migration), and UN Women, "Regional Forum on Sustainable Development, Side Event on Rural Women, Migration, and Agriculture," conference proceedings (Geneva, March 1, 2018); N. Mukhamedova and K. Wegerich, "The Feminization of Agriculture in Post-Soviet Tajikistan," *Journal of Rural Studies* 57 (2018): 128–129.

8 The *dehkan* (peasant) farms are private farms that emerged during the 1990s when former state and collective farms were dismantled and the land was allocated to member farmers. For more information, see K. Akramov and G. Shreedhar, "Economic Development, External Shocks, and Food Security in Tajikistan," IFPRI Discussion Paper 1163 (IFPRI, Washington, DC, 2012).

9 IWMI (International Water Management Institute), *Strengthening Participatory Irrigation Management in Tajikistan*, Water Policy Brief, issue 41 (Colombo, Sri Lanka: 2018).

10 FAO, "Unveiling the Matching Grants Programme for Migrants in Tajikistan," January 24, 2019, http://www.fao.org/in-action/fsn-caucasus-asia/news/news-detail/en/c/1178555/.

SOUTH ASIA

1 World Bank, World Development Indicators database, 2019.

2 FAO, "FAOSTAT Open Data," accessed October 10, 2019, http://www.fao.org/faostat/en/#data.

3 Bangladesh: Bureau of Statistics, Statistics Division, Ministry of Planning, *Report of the Household Income and Expenditure Survey 2010*; and *Final Report on Household Income and Expenditure Survey, 2016*. India: National Sample Survey Organisation, Ministry of Statistics and Programme Implementation, *Household Consumption of Various Goods and Services in India, 2004-05* (report no. 509) and *2011-12* (report no. 558); *Level and Pattern of Consumer Expenditure, 2004-05* (report no. 508); and *Key Indicators of Household Consumer Expenditure in India* (report no. NSS KI [68/1.0]). Pakistan: Bureau of Statistics, *Household Integrated Economic Survey (HIES) 2005-06; HIES 2011-12*; and *HIES 2015-16*. Sri Lanka: Department of Census and Statistics, Ministry of Policy Planning, Economic Affairs, Child, Youth and Cultural Affairs, *Household Income and Expenditure Survey 2012/13, Final Report*; and *Household Income and Expenditure Survey, 2016, Final Report*.

4 Bangladesh, National Institute of Population Research and Training, Ministry of Health and Family Welfare; Mitra and Associates; and ICF International, *Bangladesh Demographic and Health Survey 2017-18: Key Indicators* (Dhaka and Rockville, Maryland: NIPORT and ICF, 2019).

5 Bangladesh, Bureau of Statistics, Ministry of Planning, *Survey of Manufacturing Industries, 2012*; India, Ministry of Statistics and Programme Implementation, *Annual Survey of Industries Summary Results*, accessed November 2019; Nepal, Central Bureau of Statistics, *National Census of Manufacturing Establishments, 2011/12*; Pakistan, Bureau of Statistics, *Census of Manufacturing Industries, 2000-2001*; *Census of Manufacturing Industries 2005-06*; Sri Lanka, Department of Census and Statistics, "Annual Survey of Industries," final reports for 2007, 2012, and 2017.

6 World Bank, World Development Indicators database, 2019.

7 FAO, *The State of Food and Agriculture, 2019: Moving Forward on Food Loss and Waste Reduction* (Rome: 2019).

8 S. Rashid, N. Minot, and S. Lemma, "Does a Blue Revolution Help the Poor? Evidence from Bangladesh," *Agricultural Economics* 50, no. 2 (2019): 139-150; S. Rashid and X. Zhang, eds., *The Making of a Blue Revolution in Bangladesh: Enablers, Impacts, and the Path Ahead for Aquaculture* (Washington, DC: IFPRI, 2019).

9 F. Dolberg, "Review of Household Poultry Production as a Tool in Poverty Reduction with Focus on Bangladesh and India," PPLPI Working Paper No. 6 (Pro-Poor Livestock Policy Initiative, 2003); FAO, "Making Modern Poultry Markets Work for the Poor, India" (South Asia Pro Poor Livestock Policy Programme: 2012).

10 Bangladesh, *Survey of Manufacturing Industries, 2012*; India, *Annual Survey of Industries Summary Results*, accessed November 2019; Nepal, *National Census of Manufacturing Establishments, 2011/12*; Pakistan, Bureau of Statistics, *Census of Manufacturing Industries 2000-2001*; Pakistan, *Census of Manufacturing Industries, 2005-06*; Sri Lanka, "Annual Survey of Industries," final reports for 2007, 2012, and 2017.

11 M. Ravallion, "Rural Welfare Effects of Food Price Changes under Induced Wage Responses: Theory and Evidence for Bangladesh," *Oxford Economic Papers* 42, no. 3 (1990): 574-585; J. Boyce and M. Ravallion, "A Dynamic Econometric Model for Agricultural Wage Determination in Bangladesh," *Oxford Bulletin of Economics and Statistics* 53, no. 4 (1991): 361-376; X. Zhang et al., "Escalation of Real Wages in Bangladesh: Is It the Beginning of Structural Transformation?," *World Development* 64 (2014): 273-285.

12 K. Anderson, W. Martin, and D. van der Mensbrugghe, "Estimating Effects of Price-Distorting Policies Using Alternative Distortions Databases," in *Handbook of Computable General Equilibrium Modeling*, vol. 1A, ed. P. Dixon and D. Jorgenson (Oxford: North Holland, 2013); S. Rashid, R. Cummings Jr., and A. Gulati, "Grain Marketing Parastatals in Asia: Results from Six Case Studies," *World Development* 35, no. 11 (2007): 1872-1888.

13 These numbers, obtained from the World Bank's World Development Indicators database (2019), are based on surveys conducted between 2009 and 2012; coverage has likely changed since then.

14 A.U. Ahmed, *Assessing Indicators for Selection of Participants for the Vulnerable Group Development (VGD) Program in Bangladesh* (Dhaka, Bangladesh, and Washington, DC: IFPRI, 2018).

15 S. Roy et al., "Transfers, Behavior Change Communication, and Intimate Partner Violence: Postprogram Evidence from Rural Bangladesh," *Review of Economics and Statistics* 101, no. 5 (2019): 865-877.

16 A. J. Chaity and M. Al Amin, "Lead, Pesticides, Antibiotics Found in Milk," *Dhaka Tribune*, February 10, 2019; J. Wallen, "Milk Sales Plummet in Bangladesh after Studies Reveal Presence of Antibiotics and Heavy Metals," *Telegraph* (London), July 26, 2019.

17 S. Akhtar, "Food Safety Challenges—A Pakistan's Perspective," *Critical Reviews in Food Science and Nutrition* 55, no. 2 (2015): 219-226.

18 S. Baur-Yazbeck, G. Chen, and J. Roest, *The Future of G2P Payments: Expanding Customer Choice*, Consultative Group to Assist the Poor (CGAP) Focus Note (Washington, DC: CGAP, 2019).

EAST AND SOUTHEAST ASIA

1 OECD, *Economic Outlook for Southeast Asia, China and India 2019: Towards Smart Urban Transportation* (Paris: 2018).

2 V. Meadu, "Fight against Fall Armyworm in Asia Benefits from Experience in Other Regions," news features, cimmyt.org, October 31, 2019.

3 Reuters, "China's Pork Consumption Falls as African Swine Fever Spreads," May 17, 2019.

4 FAO, "ASF Situation in Asia Update," December 31, 2019, http://www.fao.org/ag/againfo/programmes/en/empres/ASF/situation_update.html.

5 China, National Bureau of Statistics, accessed December 31, 2019, http://www.stats.gov.cn/tjsj/sjjd/201910/t20191021_1704252.html.

6 FAO, "ASF Situation in Asia Update"; China, Ministry of Agriculture, "African Swine Fever Emergency Response Protocol (2019 Version)"; Chinese Academy of Agricultural Science, *Chinese Agricultural Industry Development Report 2019* (Beijing: Social Science Academic Press, 2019).

7 D. Lawder and H. Timmons, "Explainer: What a Roll-Back of Trump Tariffs on Chinese Goods May Look Like," Reuters, November 12, 2019.

8 "China-ASEAN Trade Continues to Boom amid Global Growth Slowdown, Uncertainties," *China Daily*, July 23, 2019.

9 "China Tops Foreign Buyers of Philippine Bananas, Wresting Japan's 30-yr Record," Xinhua, May 7, 2019; "China Imported More Palm Oil than It Has in Six Years," Bloomberg News, September 23, 2019.

10 R. Briones, J. P. Cruz, and R. U. Mendoza, "Rethinking Rice Policy: Philippine Rice Tariffication as a Win-Win-Win Policy for the Poor, Consumers and Farmers," May 17, 2018, available at SSRN, https://dx.doi.org/10.2139/ssrn.3179891.

11 UNICEF, *The State of the World's Children 2019: Children, Food and Nutrition: Growing Well in a Changing World* (New York: 2019); FAO, UNICEF, and WHO, *Asia and the Pacific Regional Overview of Food Security and Nutrition 2019: Placing Nutrition at the Center of Social Protection* (Bangkok: FAO, 2019); FAO, *Asia and the Pacific Regional Overview of Food Security and Nutrition 2018: Accelerating Progress towards the SDGs* (Bangkok: 2018); UNESCAP (United Nations Economic and Social Commission for Asia and the Pacific), *Inequality in Asia and the Pacific in the Era of the 2030 Agenda for Sustainable Development* (Bangkok: 2018).

12 FAO, UNICEF, and WHO, *Asia and Pacific Regional Overview*.

13 FAO, UNICEF, and WHO, *Asia and Pacific Regional Overview*.

14 Y. Wang, L. Wang, and W. Qu, "New National Data Show Alarming Increase in Obesity and Noncommunicable Chronic Diseases in China," *European Journal of Clinical Nutrition* 71, no. 1 (2017): 149.

15 WHO, "South East Asian Region: Noncommunicable Diseases," accessed January 22, 2020, https://www.who.int/nmh/ncd-tools/who-regions-south-east-asia/en/.

LATIN AMERICA AND THE CARIBBEAN

1 IMF (International Monetary Fund), "World Economic Outlook," October 2019.

2 D. Laborde and V. Piñeiro, "Trade Tensions: Implications for Latin America and the Caribbean," IFPRI and IICA Working Paper (Washington, DC: IFPRI, 2019).

3 E. Díaz-Bonilla and M. Torero, "Regional Developments: Latin America and the Caribbean," in *2017 Global Food Policy Report* (Washington, DC: IFPRI, 2017).

4 As noted, food systems also generate value added and employment in other segments of the economy—from agricultural inputs to restaurants and food outlets. Using this broader view, the authors have employed input-output matrices to estimate that food systems in Costa Rica, Jamaica, and Honduras show a participation in total GDP and employment that is about 10 percentage points higher than the indicators used in Figure 1 (even without counting further rounds of multiplier effects).

5 On average, demographic indicators in LAC regarding the sex and age structure of the population are in line with those of the world and of the aggregate of developing countries (around 1 percentage point difference). However, labor force participation in the 15–64 year old group is larger in LAC for total population, women, and youth (15–20 years old): 64 percent, 56 percent, and 50 percent, respectively, in LAC, against about 62 percent, 53 percent, and 44 percent for the world as a whole. That larger participation also leads to higher levels of recorded unemployment for total population, women, and youth compared with the world averages (7 percent, 9 percent, and 16 percent, respectively, versus 5 percent, 6 percent, and 13 percent) (data from World Bank, World Development Indicators database, 2019).

6 T. Reardon and P. Timmer, "The Economics of the Food System Revolution," *Annual Review of Resource Economics* 4 (2012): 225–264.

7 M. Nobre et al., *Atlas de las Mujeres Rurales de América Latina y el Caribe: "Al Tiempo de la Vida y los Hechos"* (Rome: FAO, 2017); A. Quisumbing et al., *Gender in Agriculture: Closing the Knowledge Gap* (Netherlands: Springer, 2014); World Bank, FAO, and IFAD, *Gender in Agriculture Sourcebook* (Washington, DC: World Bank, 2009).

8 For a review of the literature, see P. Biermayr-Jenzano, "Food Systems, Obesity, and Gender in Latin America," LAC Working Paper 1 (IFPRI, Washington, DC, 2019).

9 For information on these food-system-related health problems, see E. Díaz-Bonilla et al., *Forum on Food Systems and Obesity in Latin America and the Caribbean (LAC)* (Washington, DC: IFPRI, 2018).

10 E. Díaz-Bonilla and S. Saravia-Matus, *Los Objetivos de Desarrollo Sostenible y el Desarrollo Rural en América Latina y el Caribe: Reflexiones Sobre Costos y Financiamientos*, 2030–Alimentación, Agricultura y Desarrollo Rural en América Latina y el Caribe, no. 33 (Santiago, Chile: FAO, 2019).

Made in the USA
Las Vegas, NV
03 November 2026

57421979R00067